Data Processing Mathematics

Data Processing Mathematics

ELIZABETH BLISS
Trident Technical College

PRENTICE-HALL, INC., Englewood Cliffs, New Jersey 07632

Library of Congress Cataloging in Publication Data

Bliss, Elizabeth.
Data processing mathematics.

Inlcudes index.
1. Electronic data processing—Mathematics.
I. Title.
QA76.9.M35B55 1985 519.4 84-17744
ISBN 0-13-196155-1

Editorial/production supervision: *Karen Skrable Fortgang*
Interior design: *Chris Wolf*
Cover design: *Ben Santora*
Manufacturing buyer: *Anthony Caruso*

Printed in the United States of America

10 9 8 7 6 5 4 3 2 1

ISBN 0-13-196155-1 01

PRENTICE-HALL INTERNATIONAL, INC., *London*
PRENTICE-HALL OF AUSTRALIA PTY. LIMITED, *Sydney*
EDITORA PRENTICE-HALL DO BRASIL, LTDA., *Rio de Janeiro*
PRENTICE-HALL CANADA INC., *Toronto*
PRENTICE-HALL OF INDIA PRIVATE LIMITED, *New Delhi*
PRENTICE-HALL OF JAPAN, INC., *Tokyo*
PRENTICE-HALL OF SOUTHEAST ASIA PTE. LTD., *Singapore*
WHITEHALL BOOKS LIMITED, *Wellington*, *New Zealand*

Contents

Preface

This textbook is designed to provide data processing students with a meaningful selection of mathematical topics relevant to the field of computers. The intended audience is assumed to have a background in elementary algebra.

The first four chapters provide the basis of computer computations, logic, and design. Number base conversions are presented as algorithms. The material on logic and sets incorporate flowcharting into the presentation. The many similarities among the binary number system, logic, set notation and operations, and Boolean algebra are presented and explored.

The tie-up chapter, Chap. 5, should firmly establish the relationships that exist and demonstrate how the different approaches complement each other. It provides an excellent review of the materials in the first four chapters as well as demonstrating the close relationships that exist among the different systems.

The last five chapters in the book are relatively independent presentations. The choice of materials and the order of the presentation can be varied to best suit the needs of the related courses.

The chapter on algebra presents a review of the fundamentals with the added feature of computer notation and needs for algebraic expressions. Order of operations, use of parentheses, evaluation of expressions, and the solution of literal equations are given special emphasis to provide the necessary background for programming and data processing. The iterative process for solving polynomial equations provides the student with a new and useful approach to solutions of algebraic equations.

The basic operations for matrix algebra are presented. The subscript notation for matrices is the tool utilized for flowcharting the operations. Matrices are used as a method of solving systems of linear equations.

Patterns and general relationships are stressed throughout the text. The chapter on sequences and series uses these concepts to generate desired terms and sums. Many business applications that involve sequences and series are likewise presented.

Both the graphical and simplex methods of solution are presented in the chapter on linear programming. The exercises demonstrate a variety of business applications for linear programming.

The final chapter is a traditional coverage of logarithms. Many data processing students have missed a presentation of logarithms in previous mathematics courses. A need for some undertanding of logarithms is necessary in data processing and programming.

Every attempt is made to explain all mathematical concepts clearly and concisely. The entire thrust of the text is the use of mathematics in computer work. The purpose of the text is to present the tools of mathematics needed to understand how a computer works and to make the computer work to solve problems. The systematic presentation of procedures and methods should reinforce the systematic approach of problem solving for the student.

Many exercises are included in each section to provide the necessary practice for competence and to provide some meaningful applications of the skills. Worked-out examples precede the exercises. A systematic and logical approach to the solution of problems is stressed throughout the book.

Each chapter is preceded by an introduction providing an overview of the chapter, justification for the need for competence in the materials, and the expected outcomes from the study of the materials.

Elizabeth Bliss

Data Processing Mathematics

Different Number Bases

CHAPTER 1

1.0 INTRODUCTION

Over the years, civilizations have used various bases for their number systems. We have inherited bits and pieces of these systems and integrated them into our civilization. Why does an hour have 60 minutes, a day have two 12-hour intervals, a mile have 5280 feet? People have created the basis of numeration, and evidence suggests earlier people devised systems other than the decimal, or base 10, system in use today.

The purpose of this chapter is to provide algorithms for conversion from the decimal system to other number bases and procedures to perform arithmetic operations within these systems. A better understanding of our decimal system should result from developing skills with other number systems.

This chapter is designed to:

1. Provide a method of converting decimal numbers to other number bases.
2. Provide a method of converting any base number to decimal notation.
3. Provide a method of converting among binary, octal, and hexadecimal systems.
4. Provide the necessary arithmetic to perform addition, subtraction, multiplication, and division in binary and perform addition and multiplication in octal and hexadecimal systems.

CUNEIFORM TABLET FROM LARSA, described in 1855 by Sir Henry Rawlinson, is reproduced in part. The script is Old Babylonian cuneiform; the text is a table of square roots. Next to the top two and bottom two lines are transliterations into English script and Arabic numerals (*black*). Below the transliterations are free translations (*color*). Many such tablets were study copies made by novice scribes.

(*Courtesy of Scientific American Inc.*)

1.1 PROPERTIES OF NUMBERS

The number system utilized in today's western world has 10 as its base. Base 10 indicates there are nine different symbols plus 0, and all number values are shown as some combination of these symbols. These ten symbols are called *digits* in the decimal number system. In addition to the ten digits, each position within a number has a place value that is some power of 10.

The decimal point is the point of reference within a number. Positions to the left of the point are the whole units and the positions to the right are the fractional units of the number. The values of some of the positions are shown, both as powers of 10 and in expanded form.

					Point of reference ↓				
Position:	—	—	—	—	•	—	—	—	—
Power of ten:	10^3	10^2	10^1	10^0	•	10^{-1}	10^{-2}	10^{-3}	10^{-4}
Name:	Thousands	Hundreds	Tens	Units		Tenth	Hundredth	Thousandth	Ten thousandth
Expanded form:	1000	100	10	1	•	$\frac{1}{10}$	$\frac{1}{100}$	$\frac{1}{1000}$	$\frac{1}{10000}$

An example best illustrates the numerical system and the meaning of each position.

43,789.65 has the following value:

4 ten thousands	4×10^4	40,000
3 thousands	3×10^3	3,000
7 hundreds	7×10^2	700
8 tens	8×10^1	80
9 units	9×10^0	9
6 tenths	6×10^{-1}	.6
5 hundredths	5×10^{-2}	.05

The sum of the numbers in the right-hand column is 43,789.65, the original number.

All number systems have similar arrangements. They differ by the number of symbols available and the value of each position within the notation. The following are some of the bases with which we will be particularly concerned.

Binary. Base 2 numbers have two different symbols (0 and 1), and each position within a binary number has a place value that is some power of 2.

Octal. Base 8 numbers have eight different symbols (0, 1, 2, 3, 4, 5, 6, 7), and each position within an octal number has a place value that is some power of 8.

Hexadecimal. Base 16 numbers have sixteen different symbols (0, 1, 2, 3, 4, 5, 6, 7, 8, 9, *A*, *B*, *C*, *D*, *E*, *F*), and each position within a hexadecimal number has a place value that is some power of 16.

The number 7,654.32 as a base 8, or octal, number has the following meaning and value as a base 10 number.

7 five hundred twelves	7×8^3	3584.
6 sixty fours	6×8^2	384.
5 eights	5×8^1	40.
4 units	4×8^0	4.
3 one-eighths	3×8^{-1}	.375
2 one sixty-fourths	2×8^{-2}	.03125

The sum of the numbers in the right-hand column is 4,012.40625. This is a base 10, or decimal, number. The notation used to express this equivalence is:

$$7{,}654.32_8 = 4{,}012.40625_{10}$$

Classifying Decimal Numbers

Whole numbers (W). Whole numbers are the numbers 0, 1, 2, 3, 4, 5, . . . , where the three dots indicate that this pattern continues on without stopping.

Integers (I). The integers are the positive and negative whole numbers: . . . −3, −2, −1, 0, 1, 2, 3, 4 . . . .

Rational numbers (R). The rational numbers are numbers formed by taking the ratio of any two integers, such that the denominator is not zero. That is, a/b is a rational number if and only if $b \neq 0$.

Irrational numbers. The irrational numbers are numbers that cannot be expressed as rational numbers. $\pi, \sqrt{2}, \sqrt{3}$, and e are all examples of irrational numbers.

Real numbers ($\Re$). The real numbers include all of the above categories; $\Re$ is the set of numbers generally used in decimal arithmetic.

In computer work it is essential to know the kind of numbers being used in a program. For example, the integer value of 6.9578 is 6, and the real-number value of 6.9578 is 6.9578.

Properties of Numbers for Any Number Base

1. *Closure:* Suppose a and b are elements of a number system. If the system has operations of addition, subtraction, multiplication, and division, then $a + b$, $a - b$, $b - a$, $a \times b$, a/b, and b/a are also elements of the number system.

Example

The set of integers is closed over the operations of addition and multiplication. For instance, $(3) + (-15) = -12$; since -12 is an element of the set of integers, the integers are closed under addition. Also, $(-4) \times (6) = -24$; since -24 is an element of the set of integers, the integers are closed under multiplication.

Example

The set of whole nuumbers is closed over addition but not subtraction. $(3) + (8) = 11$ and 11 is a whole number, $(3) - (8) = -5$ and -5 is not a whole number. If an operation produces a result that is not an element of the given set, the system does not have closure for that operation.

2. *Commutative property of addition:* $a + b = b + a$;
   Commutative property of multiplication: $a \times b = b \times a$.
3. *Associative property of addition:* $a + (b + c) = (a + b) + c$;
   Associative property of multiplication: $a \times (b \times c) = (a \times b) \times c$
4. *Distributive property of multiplication over addition:* $a(b + c) = ab + ac$. This property ties the operations of addition and multiplication together.
5. *Inverse element of addition:* The additive inverse is the same element with the opposite sign. The additive inverse element of a is $-a$. Additive inverses have a sum of 0.
6. *Inverse element of multiplication:* The multiplicative inverse is the reciprocal of the given element. The multiplicative inverse element of a is $1/a$. Multiplicative inverses have a product of 1. All rational numbers have multiplicative inverses except 0.

Two identical number arrangements of digits with different bases have different numerical values. The larger the base associated with a number, the greater is the numerical value of that number.

Example

Consider 3412_8 and 3412_9.

$$3412_8 = (3 \times 8^3) + (4 \times 8^2) + (1 \times 8^1) + (2 \times 8^0)$$
$$= 1536 + 256 + 8 + 2$$
$$= 1802_{10}$$
$$3412_9 = (3 \times 9^3) + (4 \times 9^2) + (1 \times 9^1) + (2 \times 9^0)$$
$$= 2187 + 324 + 9 + 2$$
$$= 2522_{10}$$

EXERCISES

1. Give two integers that demonstrate integers do not have closure for division.
2. Give two whole numbers that demonstrate whole numbers do not have closure for subtraction.
3. Classify each of the following as real, rational, irrational, whole number, integer (each may be classified in more than one category): 3, 2.125, $\frac{2}{3}$, -8, $5\frac{2}{3}$, .123123123 . . . , $\sqrt{5}$, $\sqrt{9}$, 7.125.
4. Classify each of the following as real, rational, irrational, whole number, integer (each may be classified in more than one category): 8, -12, 3.14, $\sqrt{7}$, -6.232323 . . . , $\frac{3}{4}$, $\sqrt{16}$, $\sqrt{5.5}$.
5. If 12345 is a number whose base is less than 10, what are the possible bases for this number?
6. If 631 is a number whose base is less than 10, what are the possible bases for this number?
7. What is the value of 1234_7 in decimal notation?
8. What is the value of 1234_5 in decimal notation?
9. Suppose your mathematics instructor gives you a grade of 110 with the instructions to choose the number base you wish to use to determine your grade. The only restriction given is that the base must be less than 10. You wish to obtain the highest score possible. What base should you choose and what will be the decimal grade?
10. Suppose your mathematics instructor gives you a grade of 110 with the instructions to choose any number base you wish for the number. The grade is the number of points to be subtracted from 100. The base must be greater than 1 and you wish to receive the highest score possible. What base should you choose and what will be your score?
11. Give as many reasons as possible why 1 would be a poor selection as the base of a number system.
12. There is a small group of people interested in changing our number system to base 12, the duodecimal system. Give some reasons this group might have for its interest.

Answers

2. For example, 5 and 7 are whole numbers; $5 - 7 = -2$ and -2 is not a whole number. Therefore, whole numbers do not have closure with subtraction.
4. 8 is real, rational, whole number, integer. -12 is real, rational, integer. 3.14 is real, rational. $\sqrt{7}$ is real, irrational. $-6.232323\ldots$ is real, rational. $\frac{3}{4}$ is real, rational. $\sqrt{16}$ is real, rational, integer, whole number. $\sqrt{5.5}$ is real, irrational.
6. 631 could have a base of 7, 8, or 9.
8. $1234_5 = 125 + 50 + 15 + 4 = 194$
10. 110 to the base 2 is the smallest number possible and gives a test grade of 6. $100 - 6 = 94$.

1.2 CONVERSION FROM DECIMAL NOTATION

The following discussion provides a method of expressing a decimal or base 10 number as a number utilizing another base. The method is called an *algorithm* because it is a repetitive operation. Programming computers to perform tasks is dependent upon algorithms, which are performed accurately and quickly by computers. People tend to find algorithms tedious, laborious, and filled with opportunities for computational errors.

Algorithm for Converting Whole Decimal Numbers to Another Base

1. The whole number to be converted is the dividend of a division.
2. The desired base is the divisor.
3. The remainder resulting from the division is the rightmost bit in the output. (The word *digit* will be reserved for elements of decimal numbers and the word *bit* will be used to refer to the elements of numbers expressed in other bases.)
4. The quotient of the first division becomes the dividend of the second division, the desired base is the divisor, and the remainder from the division is the second bit (from the right) in the output.
5. Continue the division process until the quotient is 0; the remainder of that operation is the leftmost bit in the output.

Example

Convert 55 to a base 2 number.

$$2\overline{\smash{)}55}\quad \text{27 R1}$$

55 is the dividend; 2, the desired base, is the divisor. The remainder is 1. The rightmost bit in the base 2 number is 1.

$$2\overline{)27}^{\;13\ R1}$$

27, the quotient of the first division, is the new dividend, 2 is the divisor, and the remainder is 1, making the next bit (moving left) in the output 1.

$$2\overline{)13}^{\;6\ R1}$$

13 is the new dividend, 2 is the divisor, and the remainder is 1.

$$2\overline{)6}^{\;3\ R0}$$

6 is the new dividend, 2 is the divisor, and the remainder is 0.

$$2\overline{)3}^{\;1\ R1}$$

3 is the new dividend, 2 is the divisor, and the remainder is 1.

$$2\overline{)1}^{\;0\ R1}$$

1 is the new dividend, 2 is the divisor, the quotient is 0, and the remainder is 1. The process is finished; the leftmost bit of the number is 1.

Answer: The base 2 number is 110111.

A condensed version of the above process in repetitive division operations is:

2 | 55	55 is the original number to be converted; 2 is the desired base.
2 | 27 R1	27 is the quotient of the first division and the dividend of the division is 2.
2 | 13 R1	1 (top) is the remainder from the division 55 ÷ 2. This process continues six times.
2 | 6 R1	
2 | 3 R0	
2 | 1 R1	
0 R1	

The desired number consists of the remainders.

110111

This text does not intend to give rigorous developments of the algorithms suggested. However, some justification is in order to convince the user the technique suggested is logical. A concrete example shows why the algorithm is appropriate.

Factoring is the process of writing an expression as a product. For our example, the number to be factored is 55 and 2 is the desired factor.

55 is not exactly divisible by 2. However 55 can be written as 2(27) + 1.

Next, write 27 in factored form with 2 as a factor: 27 = 2(13) + 1.

Write 13 as a product of 2 and some number: 13 = 2(6) + 1.

Write 6 as a product of 2 and some number. To keep the format similar to that of the above statements, a remainder of 0 is recorded: $6 = 2(3) + 0$.

Write 3 as a product involving 2: $3 = 2(1) + 1$.

Finally, write 1 as a product involving 2: $1 = 2(0) + 1$.

Take the original statement $55 = 2(27) + 1$.

Substitute for 27 (the substitution is underlined): $2(\underline{2(13) + 1}) + 1$.

Substitute for 13 (the substitution is underlined): $2(2(\underline{2(6) + 1}) + 1) + 1$.

Substitute for 6 (the substitution is underlined): $2(2(2(\underline{2(3) + 0}) + 1) + 1) + 1$.

Substitute for 3 (the substitution is underlined): $2(2(2(2(\underline{2(1) + 1}) + 0) + 1) + 1) + 1$;

Finally, substitute for 1: $2(2(2(2(2(\underline{2(0) + 1}) + 1) + 0) + 1) + 1) + 1$.

The six factors of two are identical to the six divisions by 2 in the preceding algorithm. The six addends in the above problem are identical to the remainders from the divisions in the algorithm. The number of factors of 2 affecting each addend determines the position of the addend in the desired number. The inside addend is multiplied by 5 factors of 2, thereby occupying the position of 2^5.

$$55_{10} = 2(2(2(2(2(2(0) + 1) + 1) + 0) + 1) + 1) + 1 = 110111_2 .$$

Algorithm for Converting Positive Decimal Numbers Less than 1 to Another Base

1. The decimal number less than 1 to be converted to another base is the multiplicand of a multiplication problem.
2. The required base is the multiplier.
3. The integral part of the product is the first bit following the decimal point in the required number.
4. The integral part is subtracted from the product (resulting in a fractional quantity). The remainder becomes the new multiplicand.
5. A second product involving the new multiplicand and the desired base is performed. The integral part of the product is recorded as the next bit in the required number. The integral part is subtracted from the product.
6. The procedure is continued until the desired accuracy is obtained. Each product produces a bit in the output. Seldom can the output be expressed as an exact amount.

Example

Convert .5769 to a base 2 number with four-place accuracy. Here, .5769 is the multiplicand; 2 is the multiplier.

.5769	
× 2	
1.1538	The integer part is 1.
−1	Subtract the integral part, 1 is the first bit of the required number.
.1538	New multiplicand.
× 2	
0.3076	The integer part is 0. There is nothing to subtract; 0 is the second bit of the required number.
.3076	New multiplicand.
× 2	
0.6152	The integer part is 0. There is nothing to subtract; 0 is the third bit of the required number.
.6152	New multiplicand.
× 2	
1.2304	The integer part is 1. The desired accuracy has been attained. The fourth bit of the required number is the integral part 1.

Answer: $.5769_{10} = .1001_2$ (approximately)

There exists a close relationship between this procedure and the first algorithm presented. Multiplying by 2 is the same as dividing by $\frac{1}{2}$. Recording the carries in the multiplication by 2 is similar to recording the remainders from divisions by $\frac{1}{2}$. The procedure appears different, but in fact it is the same process as in the original algorithm.

It is not possible to use a single procedure to convert a decimal number that has both an integral and fractional part. The number must be broken into two parts. The division algorithm is used for the integral part of the number and the multiplication algorithm is used for the fractional part of the number. The two results are then combined into a single answer.

Example

Convert 1234.567 to a base 7 number with three-place accuracy.

The integral part of the decimal number is 1234. Use the division algorithm (condensed version).

7 | 1234	
7 | 176	$R_1 = 2$
7 | 25	$R_2 = 1$
7 | 3	$R_3 = 4$
0	$R_4 = 3$

$$1234_{10} = 3412_7$$

The fractional part of the decimal number is .567. Use the multiplication algorithm.

$$
\begin{array}{rl}
.567 & \\
\underline{7} & \\
3.969 & \text{Record 3 and subtract 3.} \\
.969 & \\
\underline{7} & \\
6.783 & \text{Record 6 and subtract 6.} \\
.783 & \\
\underline{7} & \\
5.481 & \text{Record 5 and subtract 5.}
\end{array}
$$

$$.567_{10} = .365_7$$

Answer: 3412.365_7

The algorithms presented are good for a number in any base. If the base is greater than 10, additional symbols are required. The letters of the alphabet are traditionally used, with A representing 10, B representing 11, and so on.

Example

Convert 2255 to a base 16 number.

$$
\begin{array}{r|l}
 & \ \ 140 \\
\hline
16 & 2255 \\
 & \underline{16} \\
 & \ \ 65 \\
 & \ \ \underline{64} \\
 & \ \ \ 15 = R_1 = F
\end{array}
\qquad
\begin{array}{r|l}
 & \ \ \ 8 \\
\hline
16 & 140 \\
 & \underline{128} \\
 & \ \ 12 = R_2 = C
\end{array}
\qquad
\begin{array}{r|l}
 & 0 \\
\hline
16 & 8 \\
 & \underline{0} \\
 & 8 = R_3
\end{array}
$$

Answer: $2255_{10} = 8CF_{16}$

EXERCISES

1. Convert each of the following to a base 2 (binary) number: 59, 271, 108, .345, .703, 3.471, 274.5. For the decimal part of the number, give as many places in the binary number as exist in the decimal number.
2. Convert each of the following to a base 8 (octal) number: 105, 356, 207, .345, .956, 17.247, 187.5. For the decimal part of the number, give as many places in the octal number as exist in the decimal number.
3. Convert each of the following to a base 16 (hexadecimal) number: 10325, 839, .1234, .9876, 87.68, 946.333. For the decimal part of the number, give as many places in the hexadecimal number as exist in the decimal number.

4. Convert each of the following as indicated. Use the same number of places as in the original number for the fractional part of the number.

$$123.45_{10} = \underline{\hspace{3cm}}_7$$
$$87.5_{10} = \underline{\hspace{3cm}}_3$$
$$13.785_{10} = \underline{\hspace{3cm}}_{16}$$

Answers

2. $105 = 151_8$; $356 = 544_8$; $207 = 317_8$; $.345 = .260_8$; $.956 = .751_8$; $17.247 = 21.176_8$; $187.5 = 273.4_8$
4. $123.45 = 234.31_7$; $87.5 = 10020.1_3$; $13.785 = D.C8F_{16}$

1.3 CONVERSION TO DECIMAL

The computer accepts numbers in decimal notation and converts these numbers into binary mode by the algorithms presented in the preceding section. All operations are performed in the binary mode, and the results are translated back to decimal notation for the user. This section presents the algorithms used for the conversion back to decimal notation. The algorithm can be used for conversion from any base to decimal notation, even though the computer uses the process only for conversion from binary to decimal.

Algorithm for Converting Integral Parts of Numbers to Decimal Notation

1. Multiply the leftmost bit by the base of the number. Add the next bit.
2. Multiply the sum from Step 1 by the base of the number. Add the next bit.
3. Repeat Step 2, using the new sum each time, until all bits have been used. The process always ends on an add operation.
4. The final sum is the desired decimal number.

Example

Convert 2453_7 to a decimal number.

1. The leftmost bit (2) is multiplied by the base 7, and the next bit is added: $(2 \times 7 + 4)$.
2. The sum (18) is multiplied by 7, and the next bit is added: $(18 \times 7 + 5)$.
3. The new sum (131) is multiplied by 7, and the next bit added: $(131 \times 7 + 3)$.

All bits have been used, the process ended with addition, and so the process is complete.

Answer: $2453_7 = 920_{10}$

The justification of this algorithm involves factoring 920 by 7.

$$920 = 7(131) + 3$$
$$131 = 7(18) + 5$$
$$18 = 7(2) + 4$$
$$2 = 7(0) + 2$$

Therefore, $920 = 7(7(7(7(0) + 2) + 4) + 5) + 3$.

Removing the grouping symbols from inside out is precisely what the above algorithm accomplishes.

$7(0) + 2 = 2$ is the original leftmost bit.

$7(2) + 4$ is identical to the leftmost bit (2) being multiplied by the base 7 and the next bit added.

$7(18) + 5$ is identical to the sum being multiplied by the base and the next bit added.

$7(131) + 2$ is the final step of the algorithm.

This is not a proof of the algorithm, but it provides justification for why it works.

Algorithm for Converting Numbers Less than One to Decimal Numbers

1. Write the bits to the right of the point as the numerator of a fraction. The denominator consists of a 1 followed by as many 0s as there are bits in the numerator.
2. Convert the numerator to a decimal number using the preceding algorithm.
3. Convert the denominator to a decimal number using the preceding algorithm.
4. To write this without a denominator, divide the numerator by the denominator, expressing the quotient to the desired accuracy.

Example

Convert $\left(\frac{20}{32}\right)_5$ to decimal notation.

$$20_5 = 10_{10} \qquad 32_5 = 17_{10}$$

Answer: $\frac{10}{17} = .588$ (three-place accuracy).

Example

Convert $.236_7$ to decimal notation.

$$.236_7 = \frac{236}{1000}_7 = \frac{125}{343}_{10} = .364$$

A number involving bits to the left and right of the decimal point must be done in two parts, with the final answer the combination of the two results.

EXERCISES

1. Convert each of the following to base 10 numbers:

$$10100110_2,\ 210102211_3,\ 76547_8,\ 11111_5,\ 8A60DB_{16}.$$

2. Convert each of the following to base 10 numbers:

$$110100111_2,\ 21212201_3,\ 34521_8,\ 101010_7,\ ABC23_{16}.$$

3. Convert each of the following to base 10 numbers (three-place accuracy):

$$\left(\frac{35}{43}\right)_8,\ \left(\frac{10101}{110111}\right)_2,\ \left(\frac{67}{CD}\right)_{16},\ .1101100_2,\ .673_8.$$

4. Convert each of the following to base 10 numbers (three-place accuracy):

$$\left(\frac{77}{234}\right)_8,\ \left(\frac{110101}{1101100}\right)_2,\ \left(\frac{CF}{2A0}\right)_{16},\ .10101001_2,\ .721_8.$$

5. Convert each of the following to base 10 numbers (two-place accuracy):

$$37.555_8,\ 101011.0011_2, A2C7.E4_{16}.$$

6. Convert each of the following to base 10 numbers (two-place accuracy):

$$73.456_8,\ 1011100.1101_2,\ 9A3F1.B4_{16}.$$

Answers

2. $110100111_2 = 423$; $21212201_3 = 1915$; $34521_8 = 14673$; $101010_7 = 17157$; $ABC23_{16} = 703523$

4. $\left(\frac{77}{234}\right)_8 = \frac{63}{156} = .403$; $\left(\frac{110101}{1101100}\right)_2 = \frac{53}{108} = .490$; $\left(\frac{CF}{2A0}\right)_{16} = \left(\frac{207}{672}\right) = .308$; $\left(\frac{10101001}{100000000}\right)_2 = \frac{169}{256} = .660$; $\left(\frac{721}{1000}\right)_8 = \frac{465}{512} = .908$

6. 59.58; 92.81; 236923.70

1.4 BINARY, OCTAL, AND HEXADECIMAL RELATIONSHIPS

The number of bits required in binary notation is very large. The three-digit decimal number 931 requires 10 bits in binary notation: 1110100011. Storage within a computer involves equipment and space; each of these involves costs. To reduce the need for large storage spaces, binary numbers are converted to numbers with larger bases. Base 8 and base 16 numbers are used because of the special relationship that exists among binary, octal, and hexadecimal numbers: $2 = 2^1$, $8 = 2^3$, $16 = 2^4$.

Decimal	Binary	Octal	Hexadecimal
0	0	0	0
1	1	1	1
2	10	2	2
3	11	3	3
4	100	4	4
5	101	5	5
6	110	6	6
7	111	7	7
8	1000	10	8
9	1001	11	9
10	1010	12	*A*
11	1011	13	*B*
12	1100	14	*C*
13	1101	15	*D*
14	1110	16	*E*
15	1111	17	*F*

The table shows that any decimal number through 7 takes up to three bits in binary, one bit in octal, and one bit in hexadecimal. The table shows that any number through 15 takes four binary bits, two octal bits, one hexadecimal bit.

When the computer accepts a decimal number, the number is converted to a binary number. If the number is stored, the binary number is converted to an octal or hexadecimal number, depending upon the system.

Algorithm for Converting from Binary to Octal

1. Starting at the point, group the binary bits in threes, moving both left and right from the point. Add zeros to the final groups to complete the threes.
2. Rename each group with an octal bit, either using the algorithm for conversion or utilizing the given chart.

Example

Convert 1101011100.01101 to octal.

oo1 101 011 100 . 011 01o The added zeros are indicated by small *o*'s

1 5 3 4 . 3 2 Octal representation of each group.

Answer: $1101011100.01101_2 = 1534.32_8$

Algorithm for Converting from Binary to Hexadecimal

1. Starting at the point, group the binary bits in fours, moving both left and right from the point. Add zeros to the final groups to make four bits.
2. Rename each group with a hexadecimal bit, either using the algorithm for conversion or utilizing the given chart.

Example

Convert 1101011100.01101 to a hexadecimal number

oo11	0101	1100	.	0110	1ooo	The added zeros are indicated by small *o*'s
3	5	*C*	.	6	8	Hexadecimal representation of each group.

Answer: $1101011100.01101_2 = 35C.68_{16}$

Algorithm for Converting from Octal to Hexadecimal

1. Convert each octal bit to three binary bits.
2. Group the binary bits into fours; the point determines the starting place.
3. Name each four-bit group as a hexadecimal bit.

Example

Convert 7516.23_8 to hexadecimal notation.

7	5	1	6	.	2	3	Octal
111	101	001	110	.	010	011	Binary
1111	0100	1110		.	0100	11oo	Regrouped binary
F	4	*E*		.	8	*C*	Hexadecimal

Answer: $7516.23_8 = F4E.8C_{16}$

Algorithm for Converting from Hexadecimal to Octal

1. Convert each hexadecimal bit to a four-bit binary number.
2. Group the binary bits into groups of three; the point determines the starting place.
3. Name each three-bit group as an octal bit.

Example

Convert $A8E2.19_{16}$ to an octal number.

A 8 E 2 . 1 9	Hexadecimal
1010 1000 1110 0010 . 0001 1001	Binary
oo1 010 100 011 100 010 . 000 110 01o	Regrouped binary
1 2 4 3 4 2 . 0 6 2	Octal

EXERCISES

1. Convert the following binary numbers to octal numbers: 10110010.11011, 10.1001, 1111.11.
2. Convert the following binary numbers to octal numbers: 11010101.0011, 110.1010, 1000.10001.
3. Convert the binary numbers in Exercise 1 to hexadecimal numbers.
4. Convert the binary numbers in Exercise 2 to hexadecimal numbers.
5. Convert the following octal numbers to hexadecimal numbers: 3761.25, 777.77, 7654.321.
6. Convert the following octal numbers to hexadecimal numbers: 6117.32, 555.22, 2460.135.
7. Convert the following hexadecimal numbers to octal numbers: $8A3.4D$, 888.88, $FDB.AC$.
8. Convert the following hexadecimal numbers to octal numbers: $9C5.E4$, 999.99, $ACE.BD$.

Answers

2. 325.14, 6.5, 10.42
4. $D5.3$, $6.A$, 8.88
6. $C4F.68$, $16D.48$, $530.2E8$
8. 4705.71, 4631.462, 5316.572

1.5 THE ARITHMETIC OF BINARY

To discover the arithmetic for binary numbers, the arithmetic in decimal numbers will be performed for the two digits and the results translated into binary.

Decimal	Binary
0 + 0 = 0	0 + 0 = 0
0 + 1 = 1	0 + 1 = 1
1 + 0 = 1	1 + 0 = 1
1 + 1 = 2	1 + 1 = 10
0 × 0 = 0	0 × 0 = 0
0 × 1 = 0	0 × 1 = 0
1 × 0 = 0	1 × 0 = 0
1 × 1 = 1	1 × 1 = 1

+	0	1
0	0	1
1	1	10

×	0	1
0	0	0
1	0	1

These number combinations are often shown in a table, as on the right. The only different result in the binary system is the sum of 1 + 1, where the number 2 must be expressed in binary notation as 10. A sum of two in binary is analogous to a sum of 10 in the decimal system—a zero is recorded in the first column and a carry of one is made to the second column.

The advantage of binary becomes more obvious when the arithmetic of numbers is investigated. For our decimal system, there are 10 sums available for each of 10 digits, making a total of 100 different sums. The commutative property cuts this number in half, but 50 different sums are necessary. Information such as this must be programmed into a computer. A like set of products must also be programmed. The 8 binary combinations described above are the only combinations necessary in the base 2 system. Regardless of how large the number is, these 8 combinations will give all possible sums and products.

Example

Addition

$$\begin{array}{r} 11011 \\ +\ 1010 \\ \hline \end{array}$$

Work from the right.

Column 1: 1 + 0 = 1

Column 2: 1 + 1 = 0 with a carry of 1

Column 3: 1 (carry) + 0 + 0 = 1

Column 4: 1 + 1 = 0 with a carry of 1

Column 5: 1 (carry) + 1 = 0 with a carry of 1

Column 6: 1 (carry) = 1

The example written in arithmetic notation is as follows:

```
 11 1      Carries
  11011
 + 1010
 100101
```

Carries frequently occur in the addition of binary numbers. There are techniques that can be used to simplify the addition. Count the number of 1s in a column. If an even number occurs the column sum is 0, with a carry of 1 for every two 1s. If the number of 1s is odd, the column sum is 1 with a carry for every two 1s in the sum.

Multiplication in the binary system of numbers resembles the multiplication in the decimal system, but the products are easier to determine.

Example

Multiply.

```
11011
X 101
```

Product 1 is 1 × 11011, which is 11011. Product 2 is 0 × 11011. This product is 0. This step is usually omitted, with the next product moved one more unit left. Product 3 is 1 × 11011, which is 11011. This multiplication written in standard form is:

```
   11011
   X 101
   11011    Product 1
  00000     Product 2, usually not written
 11011      Product 3
10000111    Sum of products
```

Subtraction is the inverse, or opposite of addition; therefore, the basic addition rules are turned around to provide the rules of subtraction.

	Addition Rules		Subtraction Rules
If	0 + 0 = 0	then	0 = 0 − 0
	1 + 0 = 1	then	1 = 1 − 0
	0 + 1 = 1	then	0 = 1 − 1
	1 + 1 = 10	then	1 = 10 − 1

The last line, 1 = 10 − 1, is handled by borrowing 1 from the column to the left. Borrowing occurs frequently in binary subtraction and is awkward to handle. Another method for subtraction will be studied in the next section.

Example

Subtract.

```
  10101
-  1010
```

Begin at the right.

Column 1: 1 - 0 = 1

Column 2: 0 - 1 = 10 - 1 = 1 with a borrow from Column 3

Column 3: 0 (because of borrow) - 0 = 0

Column 4: 0 - 1 = 10 - 1 = 1 with a borrow from Column 5

Column 5: 0 (because of borrow)

This subtraction, written in traditional form, is:

```
  1 1
  10101
-  1010
   1011
```

Division in binary arithmetic utilizes multiplication and subtraction rules. If a number is greater than or equal to the divisor, then the quotient is 1. If the number is less than the divisor, the quotient is 0. The group of bits in the dividend greater than or equal to the divisor is considered first in the division. A quotient of 1 is placed over the right most bit. Each additional bit in the dividend gives either a 0 or 1 bit in the quotient.

Example

Divide.

```
101 | 1001101
```

Step 1. 101, the divisor, will not divide into 1, 10, or 100. Therefore, the first division will involve the bits 1001 in the dividend. Put a 1 in the quotient and multiply the divisor by 1, giving 101. Then subtract: 1001 - 101 = 100.

Step 2. Bring down the next bit, 1, giving the number 1001. Put another 1 in the quotient. Multiply that by the divisor, giving the product 101. 1001 - 101 = 100.

Step 3. Bring down the next bit, 0, giving the number 1000. Put another 1 in the quotient and multiply. The product is 101. 1000 - 101 = 11.

Step 4. Bring down the next bit, 1, giving the number 111. This is greater than the divisor; therefore, put another 1 in the quotient. Multiply, record 101, and subtract, giving 10. All the bits have been used and 10 is less than the divisor 101. Therefore, the division is complete. The quotient is 1111 and the remainder is 10.

The division problem in standard division form is:

```
        1111 R10
   101 )1001101
         101
        ----
        1001
         101
         ----
         1000
          101
          ----
           111
           101
           ---
            10
```

EXERCISES

Add each of the following.

1. 10111 + 110	2. 10101 + 10010	3. 11110 + 1010	4. 1100100 + 11100

Subtract each of the following.

5. 10111 − 110	6. 10101 − 10010	7. 11110 − 1010	8. 1100100 − 11100

Multiply each of the following.

9. 10111 × 110	10. 10101 × 111	11. 11110 × 101	12. 1100100 × 100

Divide each of the following.

13. 110)10110

14. 111)10101

15. 101)11110

16. 100)1100100

Answers

2. 100111
4. 10000000
6. 11
8. 1001000
10. 10010011
12. 110010000
14. 11
16. 11001

1.6 COMPLEMENTARY ARITHMETIC

In the arithmetic of binary numbers, only positive numbers have been considered. In the section on subtraction, only problems involving subtraction of smaller positive numbers were considered. In the real world of mathematics, it is necessary to consider negative numbers, both as parts of a process and as results of an operation. Before negative numbers can be addressed, the concept of complementary arithmetic must be explored, first for decimal numbers and then for binary numbers.

A dictionary definition of a *complement* is a quantity that fills up or completes. In the base 10 numbers, the complement of 7 is 3 because 3 fills up the number 7 to the base, 10. The complement of 53 is 47. A mathematical definition of the complement of a number is the number representing the difference between the given number and the next-higher place value. The next higher placevalue for 819 is 1000. The complement of 819 is 1000 - 819, or 181. Subtractions from placevalues involve a lot of borrowing. If 1000 is renamed as 999 + 1, the complement of 819 is 999 + 1 - 819. The associative and commutative properties allow this to be rewritten as (999 - 819) + 1. The subtraction can be done without borrowing and 1 is added to the remainder: (999 - 819) + 1 = 180 + 1 = 181.

Example

Find the complement of 7539.

The next-higher placevalue is 10,000. Rewriting 10,000 as 9999 + 1, the subtraction becomes (9999 - 7539) + 1 = 2461.

Complements can be used to replace subtrahends in subtraction problems and negative numbers in any addition or subtraction problem. Extreme care must be taken to choose a placevalue higher than any number in the problem. Placevalues that are too large will never create a problem, but placevalues that are too small will produce incorrect answers. A fixed number of bits is usually assigned within a program for each number; the number exceeds any number or answer predicted within the problem.

Example

```
 12345
- 9876
------
```

Work with eight digits, which means 12345 is written as 00012345.

The complement of 9876 using eight digits is
99999999 - 9876 + 1 = 99990124.

The original problem becomes (in complement arithmetic):

```
 00012345
+99990124
---------
100002469
```

The selection of eight digits means only eight digits are recorded in the sum; the 1 at the left does not register and is known as an *overflow*. The leftmost digit (not the overflow) in the eight digits is the indicator of the sign of the arithmetic results. If the leftmost digit is 0, the answer is positive. The above example has an answer of +2469. If the leftmost digit in the eight-digit register is a 9, the answer is negative and the complement of the indicated answer.

Example

$$\begin{array}{r} 9876 \\ \underline{-12345} \end{array}$$

This obviously has a negative answer. The arithmetic of this problem in complementary arithmetic is:

$$\begin{array}{rl} 00009876 & \\ \underline{+99987655} & \quad (99999999 - 12345) + 1 \\ 99997531 & \end{array}$$

Indicates the answer is the negative of the complement of 99997531

(99999999 − 99997531) + 1 = 00002469; attach the minus sign: −2469.

Summary

1. Replace any subtrahend or negative number by its complement. The complement should be larger than any number in the problem.
2. Find the arithmetic sum of the numbers.
3. Interpret the sum. If the leftmost digit is 9, the answer is the negative of the complement of the number indicated in the solution.

Example

Find (−274) + (+741).

Step 1: (9726) + (0741)
Step 2: 0467 (The overflow 1 is not recorded.)
Step 3: +467

Note: If 726 had been used for the complement of 274, an incorrect solution would have resulted.

The process involved in complementary arithmetic is long and tedious. However, if one considers the rules involved in subtraction, borrowing, and adding signed numbers, the process may have some credibility. Electronic calculators utilize complementary arithmetic. The results in the display are the interpretations of the results of the complementary arithmetic.

EXERCISES

1. Find the complement of each of the following numbers: 7, 239, 5280.
2. Find the complement of each of the following numbers: 27, 726, 7362.

Perform each of the following subtractions by complementary arithmetic.

3. 12534	4. 2063	5. 6712	6. 75421
− 9786	− 94	− 348	−23189

Perform each of the following signed-number problems using complementary arithmetic.

7. (−908) + (+741)
8. (−908) − (+741)
9. (−741) − (908)
10. (741) − (908)

Answers

(*Note:* Circled 1s indicate an overflow.)

2. 73; 274; 2638
4. 2063 + 9906 = ①1969 = +1969
6. 75421 + 76811 = ①52232 = +52232
8. 99092 + 99259 = ①98351 = −1649
10. 741 + 99092 = 99833 = −167

1.7 COMPLEMENTARY ARITHMETIC IN BINARY NUMBERS

Section 1.6 addressed the sum of complementary numbers for decimal numbers. The concept applies to numbers in any base; the method is extremely useful for binary numbers.

Complements of Binary Numbers

The definition given in the preceding section for a complement remains the same in binary notation. The complement of a binary number is the number representing the difference between the number and the next larger placevalue.

Example

The complement of 101 is 11. The complement is 1000 − 101, or 111 + 1 − 101 = 111 − 101 + 1 = 10 + 1 = 11

Example

The complement of 1001101 is 10110011. The next-larger placevalue is 100000000, which can be written as 11111111 + 1. Subtract 1001101 and the remainder is 10110010; add 1 and the result is 10110011.

Careful analysis of this result shows that the complement of a number in binary is the opposite state of each of the bits, plus 1. In binary mode, complements are generated by replacing 0s by 1s and 1s by 0s, and then adding 1.

Example

Find the complement of 1101110010111. Replacing 1s by 0s and 0s by 1s: 0010001101000.
Add 1: 0010001101001 is the complement of 1101110010111.

Subtraction in Binary

Example

Subtraction	Complements
10111011	10111011
− 1001101	+10110011
1101110	01101110

A word of caution in complementary arithmetic in binary: work in enough bits. In computer work, bits increase in 8s—8 bits, 16 bits, and so on. In this textbook, care will be taken to work within a sufficient number, but not necessarily with 8, 16, 32, and so on.

Overflow Indicator

In computer work, a mechanism exists to indicate the problem has exceeded the constraints. This is known as an *overflow situation*, and the presence of an overflow is programmable. If the carry-in in the leftmost column of a sum is the same as the carry-out from the addition, the result is accurate. If the carry-in does not equal the carry-out, an overflow exists.

Example

8-bit arithmetic

01101101		10011111	
+11011011		+10101010	
01001000	Carry-in for left most bit is 1. Carry-out for left most bit is 1. Result valid.	01001001	Carry-in for left most bit is 0. Carry-out for left most bit is 1. Invalid.

In binary arithmetic, a 1 in the leftmost bit indicates the negative of the complement of the indicated sum.

Some examples demonstrate the overflow indicator. Use an 8-bit register.

	Base 10	Binary	Complementary form	Analysis
1.	+31	00011111	00011111	No carry-in
	- +42	-00101010	11010110	No carry-out
	-11		11110101 = -00001011	Valid
			①	
2.	-42	-00101010	11010110	Carry-in
	- +50	- +00110010	11001110	Carry-out
	-92		①10100100 = -01011100	Valid
3.	-80	-01010000	10110000	No carry-in
	+ -51	-00110011	11001101	Carry-out
	-131		①01111101	Invalid

A correct solution is possible for Example 3, if it is worked in a 10-bit register.

		①	
-80	-0001010000	1110110000	Carry-in
+ -51	-0000110011	1111001101	Carry-out
-131		①1101111101 = -0010000011	Valid

EXERCISES

Express the following decimal numbers in 8-bit binary form; negative numbers should be expressed in 2s complement.

1. 28, -28, 127, -128, -1, -75, -129
2. 35, -35, 63, -63, -2, -128, -57, -64

Perform each of the following subtractions using complementary arithmetic.

3. 00011101	4. 00010000	5. 01010010	6. 00111101
-00001011	-00001111	-00111101	-01010010

Convert each of the following to binary numbers (in 8-bit arithmetic); then perform the indicated arithmetic operations. Indicate if the solutions are valid, and convert back to decimal notation.

7. (-89) + (37)
8. (+89) + (-37)
9. (-89) - (-37)
10. (-37) - (+89)

Answers

2. 00100011, 11011101, 00111111, 11000001, 11111110, impossible, 11000111, 11000000

4.
```
  00010000
 +11110001
 ---------
  00000001
```

6.
```
  00111101
 +10101110
 ---------
  11101011
```

8.
```
  01011001
 +11011011
 ---------
  00110100   Valid, +52
```

10.
```
  11011011
 +10100111
 ---------
  10000010  Valid, -126
```

Memory Maps

Microcomputers supply maps for the memories of the instruments. A map shows where different functions are performed and the area used for the memory of the machine. The following map is copied from a Commodore reference guide.*

The numbers on the left are hexadecimal numbers that indicate the size of each location. The number is also used as the address of that location.

Memory address	
0000-3*FF*	Operating System
7*F*7-800	Screen Memory
*A*000	Sprite pointers
	Memory
	8K CBM BASIC Interpreter
	Special RAM area
*D*000	VIC-11 Register
*D*800	Color RAM
*DC*00	I/O Registers
*E*000-*FFFF*	8K CBM KERNAL OPERATING System

Suppose it has been determined that *F*12 is your present location in memory. You need to determine the amount of memory remaining.

$$\begin{array}{r} A000 \\ -\ F12 \\ \hline 90EE \end{array}$$

*Courtesy of Commodore Business Machines, Inc.

This is a hexadecimal number. If this is converted to decimal, the amount of memory remaining is 37,102 bits.

1.8 OCTAL AND HEXADECIMAL ARITHMETIC

Although all computation is performed in the binary mode in a computer, a user may need to perform octal or hexadecimal addition or subtraction to find a location within a program, to generate a graphic display, or to find the amount of memory available.

An addition table could be given for each base. However, in the real world of data processing, tables are not usually available whenever an addition task is at hand. Addition involving the meaning of octal and hexadecimal numbers is far superior. If this method is unpalatable, convert to decimal, perform the arithmetic, and then convert back to the given base.

Octal

1. There are 8 digits available - 0, 1, 2, 3, . . . , 7.
2. When the sum of a column is 8 or greater, the recorded entry in that column will be the difference between the sum and a multiple of 8 with a carry of 1 to the next column for each multiple.
3. When it is necessary to borrow in subtraction, the actual value borrowed is 8. This value is added to the existing digit.
4. If complementary arithmetic is used for subtraction, the number to generate the complement is 77 . . . 7 + 1.

Example

3456	6 + 1 = 7
+ 371	5 + 7 = 12; 12 - 8 = 4, carry 1
4047	1 + 4 + 3 = 8; 8 - 8 = 0, carry 1
	1 + 3 = 4

Example

50761	1 + 3 = 4
+67143	6 + 4 = 10; 10 - 8 = 2, carry 1
140124	1 + 7 + 1 = 9; 9 - 8 = 1, carry 1
	1 + 7 = 8; 8 - 8 = 0, carry 1
	1 + 5 + 6 = 12; 12 - 8 = 4, carry 1

Example

Subtract: 76541 - 5634.

By direct subtraction:

```
 5 13 3 9
 76541      When you borrow, you borrow
- 5634        8 units.
 ------
 70705
```

By complementary arithmetic:

```
  76541     (77777 - 5634 + 1 = 72144)
 +72144
 ------
(1)70705
```

Hexadecimal Computation

1. There are 16 digits available: 0, 1, 2, . . . , *A* (10), *B* (11), *C* (12), *D* (13), *E* (14), *F* (15). (It is wise to keep these readily available to save counting them each time.)
2. When an entry is 16 or greater, the difference between the entry and a multiple of 16 is recorded, with a carry of 1 for each multiple.
3. When it is necessary to borrow in subtraction, the actual value borrowed is 16.
4. If complementary arithmetic is used for subtraction, the number used to generate the complement is *FF* . . . *F* + 1.

Example

```
  A79DB   B(11) + 7 = 18; 18 - 16 = 2, carry 1
+  8EC7   1 + D (13) + C (12) = 26; 26 - 16 = A (10), carry 1
-------
  B08A2   1 + 9 + E (14) = 14 = 24 - 16 = 8, carry 1
          1 + 7 + 8 = 16; 16 - 16 = 0, carry 1
          1 + A (10) = B (11)
```

Example

```
                 22
           9     6   25
 A79DB    10     7    9   13   11   Indicating borrows of 16
-  8EC7   -      8   14   12    7
-------   ---------------------
           9    14   11    1    4
           9     E    B    1    4
```

The same problem can be done by complementary arithmetic:

```
                          1    1
  A79DB  →  10     7      9   13   11
+ F7139  → +15     7      1    3    9
--------   ---------------------------
            25    14     11   17   20   Indicating carries of 16
             9     E      B    1    4
```

The left most bit of 9 is not a negative indicator because the example is in hexadecimal notation.

EXERCISES

Add the following octal numbers.

1. 3571 +2164	2. 25036 + 1472	3. 6004.5 + 576.6	4. 32.56 + 7.43

Add the following hexadecimal numbers.

5. *A*468 +2*D*09	6. 179*BF* + *B*682	7. 9*A*.79*B* + *F*.*D*13	8. 48*C*.*A* +9*B*6.1

Subtract the following octal numbers.

9. 3571 -2164	10. 25036 - 1472	11. 6004.5 - 576.6	12. 32.56 - 7.43

Subtract the following hexadecimal numbers.

13. *A*468 -2*D*09	14. 179*BF* - *B*682	15. 9*A*.79*B* - *F*.*D*13	16. 9*B*6.1 -49*C*.*A*

Answers

2. 26530
4. 42.21
6. 23041
8. *E*42.*B*
10. 23344
12. 23.13
14. *C*33*D*
16. 519.7

1.9 SCIENTIFIC NOTATION

Another type of notation for large or small decimal numbers is *scientific notation.* In scientific notation, any decimal number is expressed as a product of a number between 1 and 10 (including 1) and a power of 10.

Example

$$876000 = 8.76 \times 10^5$$

$$.000076 = 7.6 \times 10^{-5}$$

General Procedure for Expressing a Decimal Number in Scientific Notation

1. Place the decimal point between the two leftmost significant digits. This creates a number greater than or equal to 1 and less than 10.

2. The exponent of 10 is the number of places the existing decimal point is moved to its new position. If the original number is less than 1, the exponent is negative. If the original number is greater than 10, the exponent is positive. If the decimal point is not moved, the exponent is 0.

Scientific notation is most often used in the scientific community. It is a notation that allows the magnitude of quantities to be read and understood quickly.

Examples

The distance to the sun is 93,000,000 miles = 9.3×10^7 miles.

The speed of light is 186,000 = 1.86×10^5 miles/sec.

The mass of an electron is .00000000000000000000000000000201 Kg = 2.01×10^{-30} Kg.

Performing arithmetic operations on numbers expressed in scientific notation requires careful observation of the laws of exponents.

Addition and Subtraction

Quantities expressed in scientific notation can be added or subtracted only if the exponents of 10 are equal.

Example

$$(3.45 \times 10^7) + (6.2 \times 10^7) = 9.65 \times 10^7$$

$$(4.87 \times 10^{-4}) + (8.01 \times 10^{-4}) = 12.88 \times 10^{-4}$$

The latter number is no longer in scientific notation. To write the sum in scientific notation, change 12.88 to 1.288 and increase the exponent by 1: 1.288×10^{-3}.

It is not possible to add $(3.45 \times 10^8) + (3.47 \times 10^{10})$ as written. To add, change the number with the smaller exponent to a number with the greater exponent. Change 3.45×10^8, which has the smaller exponent, to $.0345 \times 10^{10}$. Now, $.0345 \times 10^{10}$ and 3.47×10^{10} can be added; their sum is 3.5045×10^{10}.

Multiplication

Any two numbers expressed in scientific notation can be multiplied by multiplying the decimal parts and adding the exponents of 10.

Example

$$(3.45 \times 10^7) \times (2.0 \times 10^4) = 6.90 \times 10^{11}$$

$$(4.2 \times 10^{-4}) \times (8.0 \times 10^6) = 33.6 \times 10^2$$

The latter number is not scientific notation and should be rewritten as 3.36×10^3.

$$(2.0 \times 10^{-6}) \times (3.6 \times 10^{-10}) = 7.2 \times 10^{-16}$$

In each case, the number parts are multiplied and the exponents are added.

Division

Any two numbers expressed in scientific notation can be divided by dividing the decimal parts and subtracting the exponent in the divisor from the exponent in the dividend.

Example

$$(3.45 \times 10^{6})/(5.0 \times 10^{10}) = .69 \times 10^{-4}$$

This number is not in scientific notation and should be renamed as 6.9×10^{-5}.

$$(4.84 \times 10^{8})/(4.4 \times 10^{-6}) = 1.1 \times 10^{14}$$

Powers

Numbers written in scientific notation can be raised to powers by raising the decimal part of the number to the power and multiplying the exponent of 10 by the power.

Example

$$(2.00 \times 10^{3})^4 = (2.00)^4 \times (10^3)^4 = 16.0 \times 10^{12} = 1.6 \times 10^{13}$$

$$(4.0 \times 10^{-2})^4 = (4.0)^4 \times (10^{-2})^4 = 256.0 \times 10^{-8} = 2.56 \times 10^{-6}$$

Scientific notation is useful notation for decimal numbers that are very large or very small. Computation involving these types of numbers is best performed in scientific notation.

Example

Use scientific notation to compute

$$\frac{840{,}000{,}000 \times .00012}{60000000}$$

Using scientific notation, this becomes $(8.4 \times 10^8) \times (1.2 \times 10^{-4})/(6.0 \times 10^7)$.

$$8.4 \times 1.2/6.0 = 1.68$$

$$8 + (-4) - 7 = -3$$

Answer: 1.68×10^{-3}

Often numbers are translated to scientific notation to perform computation, but the answer needs to be rewritten in decimal notation. To convert a number written in scientific notation to decimal notation, move the decimal point as many places as the exponent of 10 indicates—right if the exponent is positive and left if the exponent is negative.

Example

$$8.03 \times 10^4 = 80{,}300$$

$$6.60 \times 10^{-6} = .00000660$$

Scientific notation is useful for approximating a solution or predicting the relative size of a result.

Example

Approximate $89{,}678 \times .000000198/9800000$. Approximate this problem as $(9 \times 10^4) \times (2 \times 10^{-7})/(1.0 \times 10^7)$. An approximate answer is then 18×10^{-10}, or .0000000018. This problem is written as $89{,}678 \times .000000198/9800000 \approx .0000000018$. $\approx$ is the mathematical symbol for approximates.

Approximating the expected size of answers is an important use of scientific notation. Errors in entering data can be detected by the individual who has some idea of the size of the expected result.

EXERCISES

Express each number in scientific notation.

1. 32,145
2. 867,000
3. 4,600,000
4. .234
5. .0005006
6. .000000078
7. 234.1
8. 189×10^3
9. $.000231 \times 10^4$
10. $.95 \times 10^{-2}$

Perform the indicated operation in each case. Leave your answer in scientific notation.

11. $(2.35 \times 10^3) + (7.0 \times 10^3)$
12. $(7.95 \times 10^{-4}) + (1.05 \times 10^{-4})$
13. $(6.8 \times 10^6) + (9.004 \times 10^6)$
14. $(2.003 \times 10^{-1}) + (9.9 \times 10^{-1})$
15. $(1.6 \times 10^3) + (2.7 \times 10^4)$
16. $(6.81 \times 10^{-2}) + (5.0 \times 10^{-3})$

17. $(2.35 \times 10^{6}) \times (4.0 \times 10^{4})$
18. $(3.2 \times 10^{-2}) \times (2.0 \times 10^{-5})$
19. $(8.01 \times 10^{3}) \times (9.3 \times 10^{-5})$
20. $(4.4 \times 10^{-6}) \times (2.7 \times 10^{7})$
21. $(3.41 \times 10^{2})/(1.63 \times 10^{4})$
22. $(8.4 \times 10^{6})/(4.2 \times 10^{-2})$
23. $(1.56 \times 10^{-4})/(8.96 \times 10^{6})$
24. $(4.84 \times 10^{3})/(9.68 \times 10^{3})$
25. $(3.0 \times 10^{4})^{3}$
26. $(2.4 \times 10^{-2})^{4}$

Change the notation to scientific notation, perform the calculations, and convert the answer back to decimal notation.

27. $[8{,}400{,}000 \times .00002 \times (120)^{3}]/.00084$
28. $[.123 \times (3400)^{2}]/12{,}000{,}000$
29. $(93{,}000{,}000 \times .002)/(1200 \times .000004)$
30. $[(.00008)^{2} \times .23]/.00056$
31. $(270 \times 60{,}000)/.36$
32. $(.00048 \times 2{,}000)/23{,}000{,}000$

Use scientific notation to estimate an answer to each of the computational problems.

33. $(987{,}659 \times 569)/.0002$
34. $.000000008 \times 678{,}000{,}000 \times (.02)^{4}$
35. $(2{,}000{,}000 \times .000008)/.004$
36. $(240{,}000 \times 60 \times 60)/.0005$
37. $[7{,}890{,}000 \times .0123 \times (98)^{3}]/.08$
38. $(.00000456 \times .009)/(.003)$

Answers

2. 8.67×10^{5}
4. 2.34×10^{-1}
6. 7.8×10^{-8}
8. 1.89×10^{5}
10. 9.5×10^{-3}
12. 9.00×10^{-4}
14. 1.1903×10^{0}
16. 7.31×10^{-2}
18. 6.4×10^{-7}
20. 1.188×10^{2}
22. 2.0×10^{8}
24. 5.0×10^{-1}

26. 3.31776×10^{-7}
28. $[1.23 \times 10^{-1} \times (3.4 \times 10^3)^2]/(1.2 \times 10^7) = 11.849 \times 10^{-2} = .11849$
30. $[(8 \times 10^{-5})^2 \times 2.3 \times 10^{-1}]/(5.6 \times 10^{-4}) = 26.285714 \times 10^{-7} = .0000026185714$
32. $(4.8 \times 10^{-4} \times 2 \times 10^3)/(2.3 \times 10^7) = 4.173913 \times 10^{-8} = .00000004173913$
34. $8 \times 10^{-9} \times 7 \times 10^8 \times 16 \times 10^{-8} \approx 1{,}000 \times 10^{-9} \approx .000001$
36. $2.5 \times 10^5 \times 6 \times 10^1 \times 6 \times 10^1/5 \times 10^{-4} \approx 18 \times 10^{11} \approx 1800000000000.$
38. $5 \times 10^{-6} \times 1 \times 10^{-2}/3 \times 10^{-3} \approx 2 \times 10^{-5} \approx .00002$

1.10 GRAPHIC DESIGNS

Characters can be designed for a screen display by the use of binary notation. An 8 X 8 matrix is used to design characters. The column headings are 0 through 7 from right to left. Each number represents the exponent of 2 for the column. The rows are numbered 0 through 7 from the top to the bottom.

The filled-in squares form the desired character, as shown in the figure. The binary numbers on the right are formed by 0s representing empty squares and 1s representing the filled-in squares. The binary numbers are then converted to decimal notation.

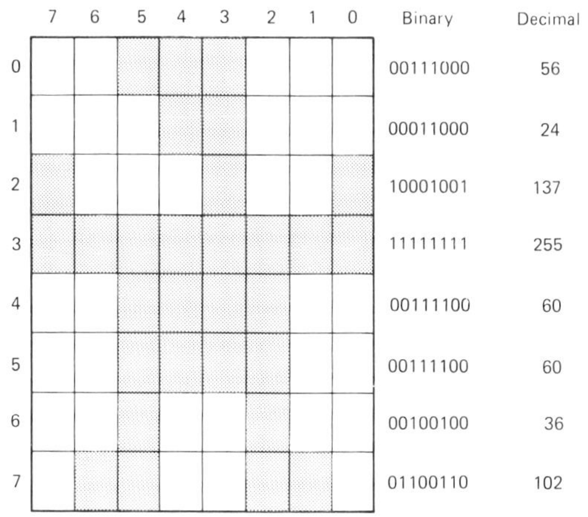

Mr. Square Man

These decimal numbers are used as data in the program that generates Mr. Square Man. The computer data statement for the character is:

Data 56, 24, 137, 255, 60, 60, 36, 102

The character can be animated by moving the numbers in the data statement in a consistent pattern, generating additional 8 X 8 grids.

CHAPTER 1 REVIEW EXERCISES

Convert each of the following decimal numbers to the indicated base (same accuracy as the original number).

1. $29 = \_\_\_\_\_\__2$
2. $10325 = \_\_\_\_\_\__{16}$
3. $.864 = \_\_\_\_\_\__2$
4. $103.906 = \_\_\_\_\_\__8$

Convert each of the following numbers to decimal.

5. 101101101_2
6. 7146_8
7. $.101101_2$
8. $67.AC_{16}$

Convert each of the following as indicated.

9. $AD9_{16} = \_\_\_\_\_\__2$
10. $11011.101_2 = \_\_\_\_\_\__8$
11. $76.35_8 = \_\_\_\_\_\__2$
12. $A58.F9_{16} = \_\_\_\_\_\__8$

Add each of the following.

13. $\begin{array}{r} 101101_2 \\ +\ 11101_2 \\ \hline \end{array}$
14. $\begin{array}{r} 7654_8 \\ +1352_8 \\ \hline \end{array}$
15. $\begin{array}{r} A9D4.F_{16} \\ +\ 165.7_{16} \\ \hline \end{array}$

Subtract each of the following.

16. $\begin{array}{r} 101101_2 \\ -\ 11101_2 \\ \hline \end{array}$
17. $\begin{array}{r} 7654_8 \\ -1352_8 \\ \hline \end{array}$
18. $\begin{array}{r} A9D4.F_{16} \\ -\ 165.7_{16} \\ \hline \end{array}$

Multiply.

19. $\begin{array}{r} 11011 \\ \times\ 101 \\ \hline \end{array}$

Divide.

20. $101\overline{)11011}$

Do the following problems in binary 8-bit arithmetic, using addition and complementary arithmetic. Show the answers as plus or minus a binary number and convert the answer to decimal. Indicate the validity of the answer.

21. (+37) - (+105)
22. (-37) + (+105)
23. (-37) - (-105)
24. (-37) - (+105)
25. (+37) + (+105)

CHALLENGE PROBLEMS

26. A man goes to the bank and withdraws \$1000. He gives the teller 10 bags and asks that the money be distributed in the bags so he can use one or more bags to pay an individual any dollar amount less than \$1000. How did the teller distribute the money in the bags?
27. In a game called "Twenty Questions," a contestant was asked to determine the number written on a piece of paper. The number was less than 500,000. What would the 20 questions be that would determine the number? The rules of the game allow only yes and no answers to questions.

Chapter 1 Review Answers

2. 2855
4. 67.717
6. 3686
8. 103.671
10. 33.5
12. 5130.762
14. 11226
16. 10000_2
18. $A86F.8_{16}$
20. 101 R10
22.
```
    11011011
    01101001        Carry-in = carry-out = 1; valid.
  1|01000100 = +68
```
24.
```
    11011011
    10010111        Carry-in ≠ carry-out; invalid.
  1|01110010
```

Sets

CHAPTER 2

2.0 INTRODUCTION

The computer revolution allows the processing and storage of vast quantities of information. Using the information requires classification of the information into groups (called *sets* in mathematics) with common characteristics. Operations are then performed on groups of data rather than on individual bits. It is necessary to know the operations available, the outcomes that can be expected, and the algebra of sets.

People who are concerned with specific elements but never look for the pattern of the whole picture are often described as people who cannot see the forest for the trees. A good manager needs to put elements into the whole picture. Long-range planning, behavioral objectives, and goals can help accomplish this result.

Computers assist people in seeing the whole picture. Information is readily available, programs allow for simulation of plans of action and outcomes, and records show similiar situations and reactions.

This chapter deals with the vocabulary, operations, and graphic representations of sets.

This chapter is designed to:

1. Provide the vocabulary of sets, the operations of sets, and the means of translating verbal problems to set notation.
2. Provide the fundamental concepts of Venn diagrams, the graphic representation of sets.

3. Provide the concepts needed to utilize Venn diagrams in determining the nature and number of elements within sets and subsets.

2.1 THE VOCABULARY OF SETS

A set is a group or collection of elements. Braces, { }, are used to denote sets. A set must be defined so it can be determined if an element is a member of a set or not. This is done by listing the elements in the braces or by giving a definition that determines the elements. The order that elements are listed is of no importance.

Example

$\{a, b, d, e, c\}$ and {the first five letters of the alphabet} define the same set.

The vocabulary of sets is essential before sets can be operated on and used.

Equal Sets

Two sets are *equal* if the two sets contain the same elements and the same number of elements.

Example

$A = \{a, c, e\}$ and $B = \{e, a, c\}$; $A = B$

Equivalent Sets

Two sets that have the same number of elements are *equivalent*.

Example

$A = \{1, 2, 3, \ldots, 7\}$ and $B =$ {days of the week}; A is equivalent to B.

Null Set

The null set is an empty set, the set with no elements. The empty set is denoted by $\emptyset$ or { }.

Example

{living Civil War veterans} $= \emptyset$

Universal Set

The *universal set* is the entire collection of things under consideration. The symbol is $\mathcal{U}$.

Example

The universal set for the set of odd integers is the set of all integers.

Infinite Set

An *infinite set* is a set that contains an infinite number of elements.

Example

The set of whole numbers, $\{0, 1, 2, 3, \ldots\}$, is an infinite set. The three dots indicate that the pattern continues without ending.

Subset

A subset is a set of elements that is completely contained in another set. If A is a proper subset of B ($A \subset B$), then A and B are not equal. The symbol $\subseteq$ means a set is either a proper subset or equal to another set. A set of n elements has 2^n possible subsets.

Example

Given $A = \{0, 1, 2\}$, $B = \{0, 1, 2, \ldots, 10\}$ and $C = \{$integers between 0 and 10, inclusive$\}$. Then $A \subset B$ and $B \subseteq C$. Also, $A \subset C$ and $C \subseteq B$. Also, A has $2^3 = 8$ subsets.

EXERCISES

Write each set by using braces and listing the elements.

1. The whole numbers less than 8
2. The integers four units or less from 1
3. The odd whole numbers
4. The football players living on the moon

Given that $A = \{$the months of the year$\}$. Classify each of the following as *true* or *false*. If the statement is false, make it true, ($\in$ is the symbol for "is an element of" and $\notin$ is the symbol for "is not an element of.")

5. May $\in A$
6. Tuesday $\in A$
7. Spring $\notin A$
8. $\emptyset \subset A$
9. $\emptyset \in A$
10. $\{$March, May$\} \subset A$
11. A is equivalent to $\{1, 2, 3, \ldots, 12\}$
12. $\{$March, Monday, May$\} \subseteq A$

Classify the first set in each case as equal to, equivalent to, a proper subset of ($\subset$), or a subset or equal to ($\subseteq$) the second set.

13. $\{a, b, c\}$ and $\{c, a, b\}$
14. {@, #, &, *} and $\{1, 2, 3, 4\}$
15. $\{1, 2, 3\}$ and {integers}
16. $\{2, 4, 6, 8\}$ and $\{1, 2, 3, 4, 5, 6, 7\}$

Given $A \subset B$, $x \in A$, and $y \in B$.

17. Must $x \in B$ be true?
18. Must $y \in A$ be true?

Given $A = \{1, 2, 3\}$

19. How many subsets in A?
20. List the subsets of A?

Answers

2. $\{-3, -2, -1, 0, 1, 2, 3, 4, 5\}$
4. $\{\ \}$ or $\emptyset$
6. Tuesday $\notin A$
8. True
10. True
12. {March, Monday, May} $\not\subseteq A$
14. Equivalent
16. None
18. No
20. $\emptyset, \{1\}, \{2\}, \{3\}, \{1, 2\}, \{2, 3\}, \{1, 3\}, \{1, 2, 3\}$

2.2 THE ALGEBRA OF SETS

There are three operations associated with sets. These operations are defined and the notation associated with each operation is presented.

Union, $\cup$. The *union* of two sets is the connective creating a set containing all elements in either or both of the given sets.

$$A \subseteq (A \cup B) \qquad B \subseteq (A \cup B) \qquad A \cup B \subseteq \mathfrak{U}$$

Example

$$A = \{1, 2, 3\}, B = \{2, 3, 4\}, A \cup B = \{1, 2, 3, 4\}$$

The word *or* in verbal statements translates to union in set notation.

Example

Candidates must be over 35 years of age or have had at least 5 years computer programming experience.

$A = \{\text{Individual over 35 years of age}\}$

$B = \{\text{Individuals with at least 5 years computer programming experience}\}$

Candidates $= A \cup B$

Intersection, $\cap$. The *intersection* of two sets is the connective creating the set of the common elements of two sets. The intersection of two sets is usually a smaller set than either of the original sets.

$$(A \cap B) \subseteq A \qquad (A \cap B) \subseteq B$$

Example

$A = \{1, 2, 3\}, B = \{2, 3, 4\}, A \cap B = \{2, 3\}$

The words *and* and *but* in verbal statements, translate to intersection in set notation. *And* is often implied in verbal statements, not written.

Example

Women over 65 years of age.

$A = \{\text{people over 65}\}, B = \{\text{women}\}, A \cap B =$ women over 65 years of age

Complement, $'$. The *complement* of a set is the connective creating the set of elements in the universal set not in the given set. It is essential to know the population of the universal set in order to determine the complement of a set.

$$A \cup A' = \mathcal{U} \qquad A \cap A' = \phi$$

Example

$A = \{2, 3, 4, 5, 6\}, \mathcal{U} = \{1, 2, 3, 4, 5, 6, 7, 8, 9\}, A' = \{1, 7, 8, 9\}$

The word *not* in verbal statements translates to complement.

Example

$\mathcal{U} = \{\text{all citizens of the United States}\}, A = \{\text{all male citizens of the United States}\},$
$A' = \{\text{all female citizens of the United States}\}$

The Algebra of Sets

Chapter 1, listed properties that pertained to all numbers. A similiar list of properties is given for sets. These properties hold for all sets.

1. Commutative properties: $A \cup B = B \cup A$; $A \cap B = B \cap \mathrm{A}$.
2. Associate properties: $A \cup (B \cup C) = (A \cup B) \cup C$, $A \cap (B \cap C) = (A \cap B) \cap C$
3. Distributive properties: $A \cup (B \cap C) = (A \cup B) \cap (A \cup C)$; $A \cap (B \cup C) = (A \cap B) \cup (A \cap C)$

Example

$$\mathfrak{U} = \{1, 2, 3, 4, \ldots, 10\}$$

$$A = \{2, 4, 6, 8, 10\}$$

$$B = \{1, 3, 5\}$$

$C = \{1, 2, 3, 4, 5, 6\}$	$A \cap B = \phi$
$A \cup B = \{1, 2, 3, 4, 5, 6, 8, 10\}$	$A' = \{1, 3, 5, 7, 9\}$
$A \cup C = \{1, 2, 3, 4, 5, 6, 8, 10\}$	$A \cap C = \{2, 4, 6\}$
$B \cup C = \{1, 2, 3, 4, 5, 6\}$	$B \cap C = \{1, 3, 5\}$
$B' = \{2, 4, 6, 7, 8, 9, 10\}$	$C' = \{7, 8, 9, 10\}$
$B' \cap C' = \{7, 8, 9, 10\}$	$A \cap (B \cup C) = \{2, 4, 6\}$

Example

The employees of JARO Inc. are classified as salaried (S), female (F), office (O), and manufacturing (M).
Write the set notation for each.

1. Salaried male office workers. Although *and* is not written, the meaning of the collection of words is salaried *and* male *and* office.

$$S \cap F' \cap O$$

2. Not salaried or a female office employee.

$$S' \cup (F \cap O)$$

3. Salaried workers who are not involved in manufacturing.

$$S \cap M'$$

4. Male manufacturing employees or female employees not in the office.

$$(F' \cap M) \cup (F \cap O')$$

Verbal problems always seem to be the more difficult to translate. If a set is becoming more restrictive, the connective will be intersection. The word *or* must appear in the statement for a union. The word *and* is often implied.

EXERCISES

Given $\mathcal{U} = \{1, 2, 3, 4\}, A = \{1, 4\}, B = \{1, 3\}, C = \{2, 4\}$
Write the set representation of each of the following.

1. $A \cup B$
2. $A' \cup C$
3. $A \cap C$
4. $B \cap C'$
5. $B' \cup C'$
6. $A \cap (B \cup C)$
7. $(A \cup B)'$
8. $A' \cup B'$
9. $A' \cap B'$

Given

$\mathcal{U} = \{\text{student body at your school}\}$ $A = \{\text{freshman students}\}$

$B = \{\text{female students}\}$ $C = \{\text{Data processing students}\}$

Give a verbal description of each of the following sets.

10. $A \cup C$
11. $B \cap C$
12. $A' \cap B'$
13. $B \cup (A' \cap C)$
14. $(A \cup C)'$
15. $A' \cap C'$
16. $A \cap (B' \cup C)$

Given:

$\mathcal{U} = \{\text{all customers of A. B. Smith Co.}\}$ $A = \{\text{customers within a 20-mile radius}\}$

$B = \{\text{customers over 50 years of age}\}$ $C = \{\text{customers who own a color television}\}$

$D = \{\text{customers with a charge account}\}$

Give the set notation for each of the following.

17. Customers who live within a 20-mile radius of the store who charged a color television.
18. Customers 50 years of age or younger who have a charge account.
19. Customers who are over 50 years of age or who do not have a color TV.
20. Younger customers (up to 50) who live more than 20 miles from the store or customers who do not have a charge account.

Answers

2. $\{2, 3, 4\}$
4. $\{1, 3\}$
6. $\{1, 4\}$
8. $\{2, 3, 4\}$
10. Freshman or data processing students
12. Upperclassmen male students
14. Not (freshmen or data processing students) = upperclassmen who are not data processing students
16. Freshman male students or freshman data processing students
18. $B' \cap D$
20. $(B' \cap A') \cup D'$

2.3 VENN DIAGRAMS

To help in determining the population of a set, a pictorial description of the sets, called a *Venn diagram*, can be used. In a Venn diagram, the following representations are made:

The universal set is represented by a rectangle.

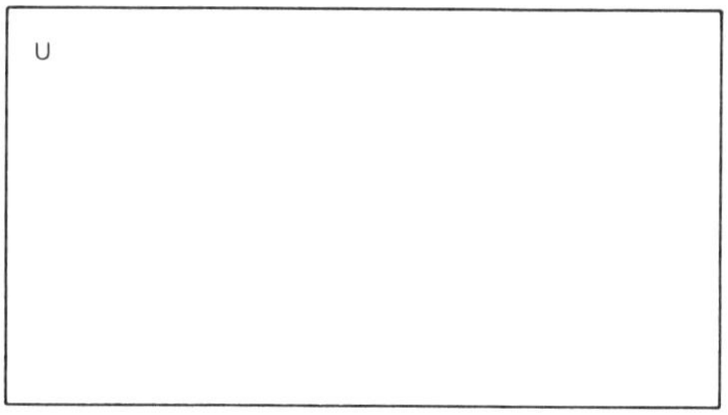

A subset is shown by a circle within the rectangle. The sizes of the rectangle and circle do not indicate the relative sizes of the sets represented.

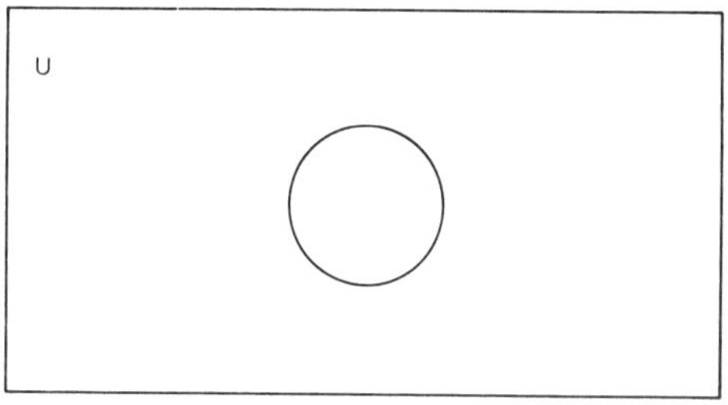

Subset A is represented by a shaded circle within a rectangle.

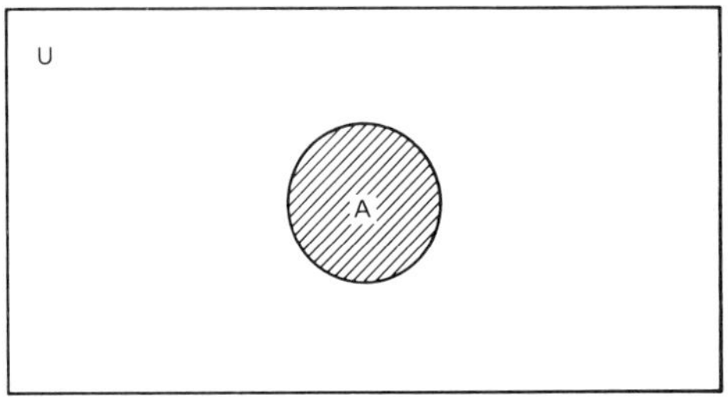

If there is more than one subset of $\mathcal{U}$, each subset is represented by a circle. The relationship existing between the subsets determines the positions of the circles within the rectangle.

All A are B, or $A \subset B$.

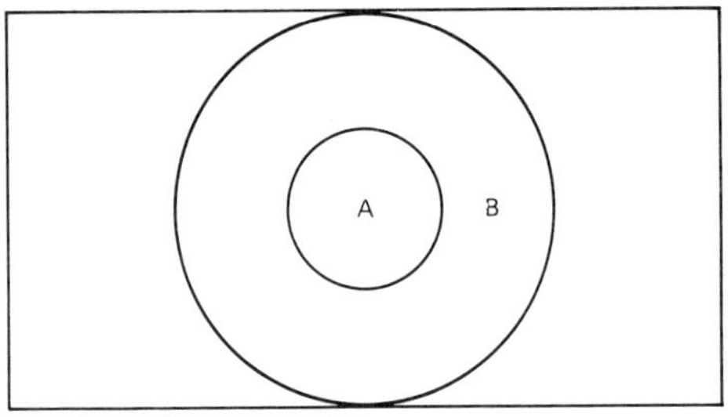

No A are B, or $A \cap B = \phi$.

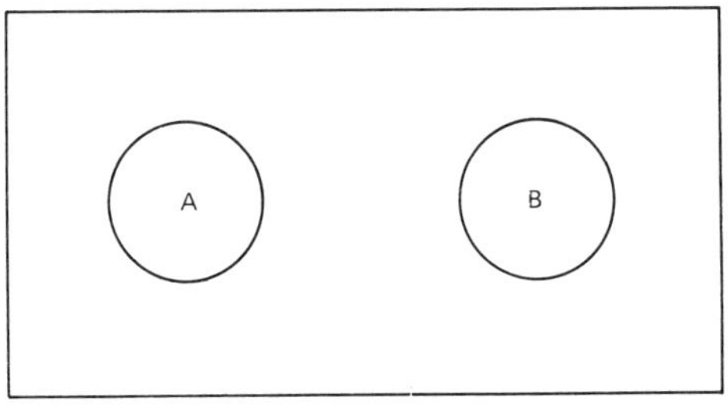

Some A are B, or $A \cap B \neq \phi$.

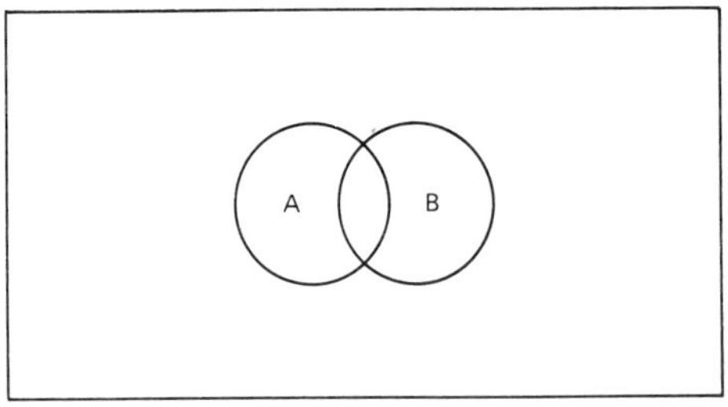

Shading is used to represent the areas denoted by the connectives union, intersect, and complement.

A or $B = A \cup B$ (shade A combined with B).

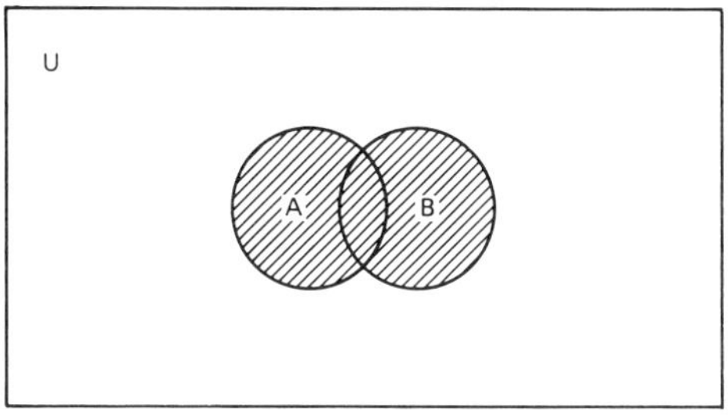

A and $B = A \cap B$ (shade A shared with B).

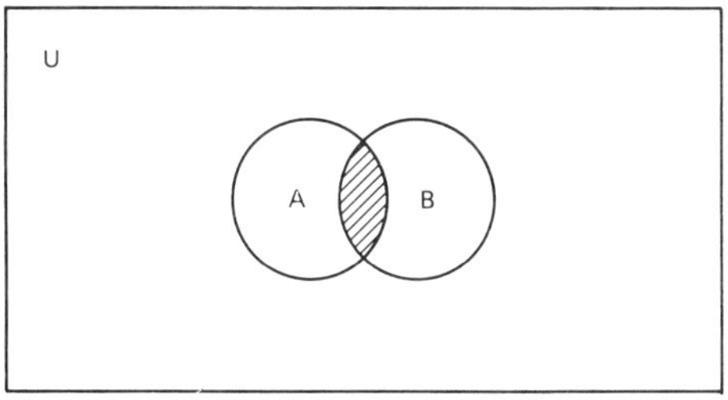

not $A = A'$ (shade all except A).

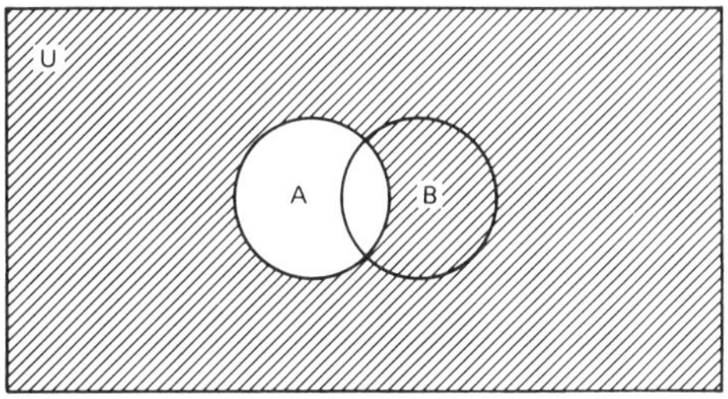

In a two-subset Venn diagram, there are basically four areas besides the universal set. The given diagrams show the areas and the descriptions of the sets.

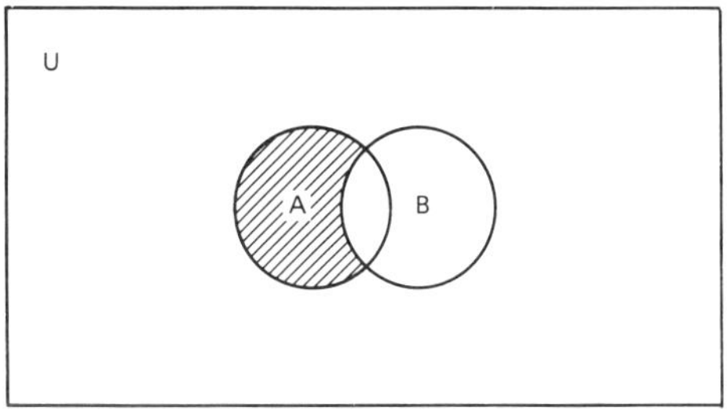

$A \cap B'$

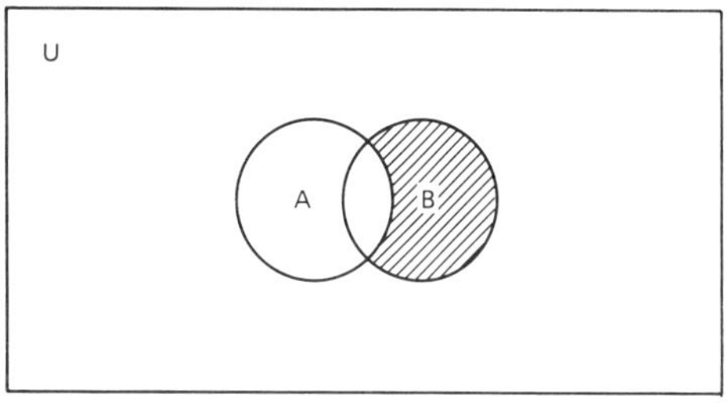

$A' \cap B$

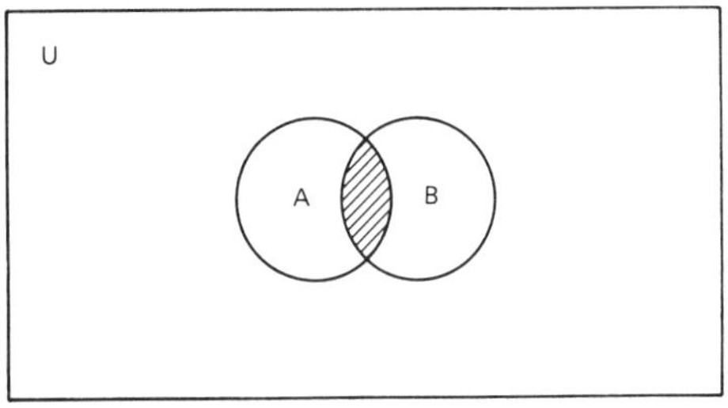

A ∩ B

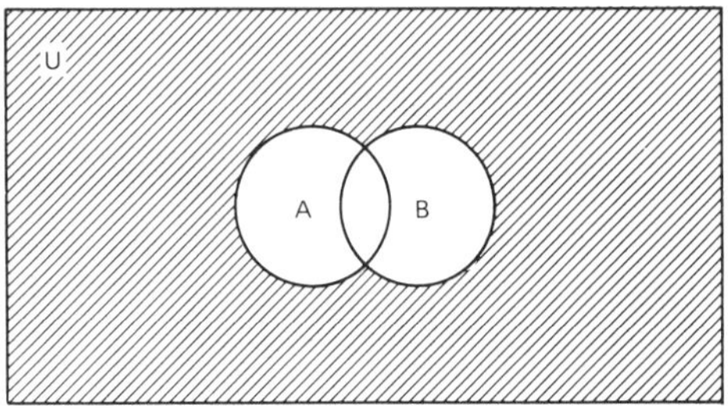

(A ∪ B)′ = A′ ∩ B′

In a three-subset Venn diagram, there are eight areas, as numbered in the diagram. (An area does not cross a boundary line)

Area 1 is in A, in B, and in C: $A \cap B \cap C$.

Area 2 is in B and in C but not in A: $A' \cap B \cap C$.

Area 5 is in B but not in A and not in C: $A' \cap B \cap C'$.

Area 8 is not in A, not in B, and not in C: $A' \cap B' \cap C'$.

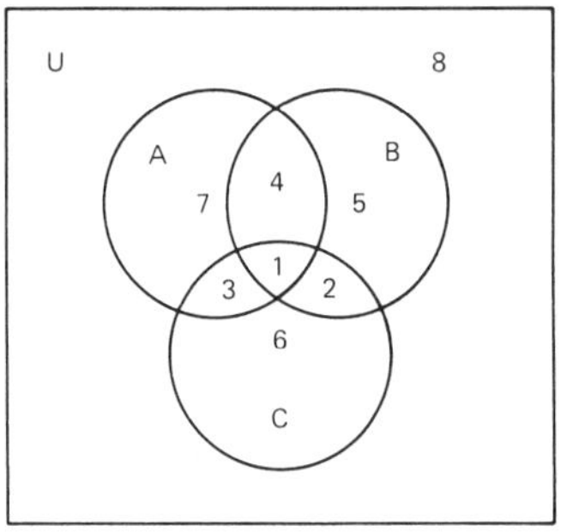

The naming of areas 3, 4, 6, and 7 is left as an exercise. Basically, all areas of a two- or three-subset Venn diagram have been named. If a picture is presented, the set notation that represents the picture can be written.

Example

Represent $(A' \cup B) \cap (A' \cup C)$ in a Venn diagram.

There are three subsets involved, so there are three circles within the rectangle. First do the two parentheses (separately).

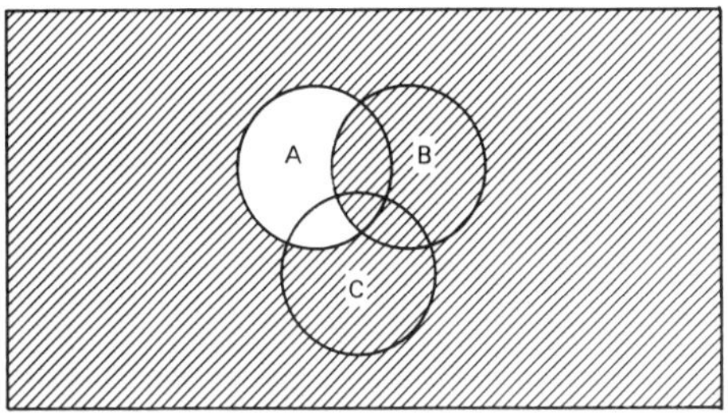

A′ ∪ B

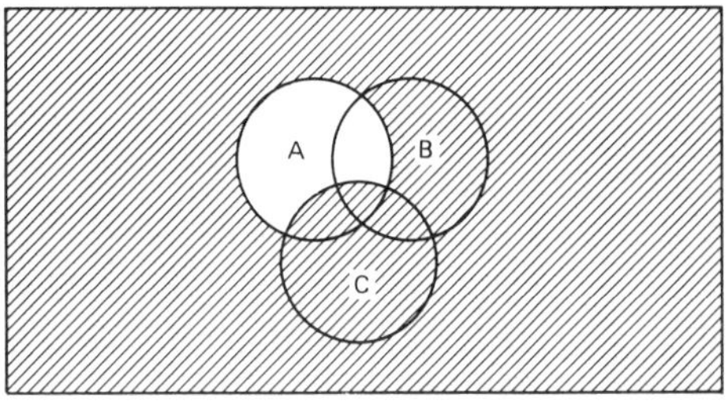

A′ ∪ C

Because these two parentheses are connected with $\cap$, the final result is the areas shaded in both the above diagrams.

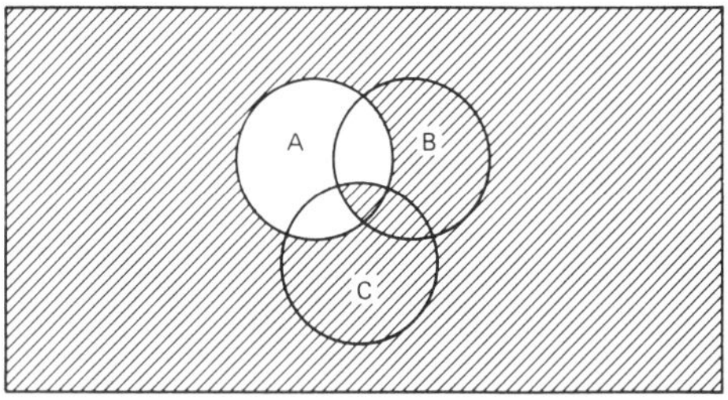

$(A' \cup B) \cap (A' \cup C)$

EXERCISES

Draw a Venn diagram to represent each of the following sets.

1. $A \cup B'$
2. $A' \cup B$
3. $A \cup B \cup C'$
4. $A \cup B' \cup C$
5. $A \cap B'$
6. $A' \cap B$
7. $A \cap B \cap C'$
8. $A \cap B' \cap C$
9. $(A \cap B) \cup (A \cap C)$
10. $(A \cap B) \cup (B \cap C)$

Given:

$$\mathfrak{U} = \{0, 1, 2, 3, 4, 5, 6, 7, 8, 9, 10\}$$
$$A = \{2, 4, 6, 8, 10\}$$
$$B = \{1, 3, 5, 7, 9\}$$
$$C = \{0, 1, 2, 3, 4, 5\}$$

11. Draw a Venn diagram to represent sets A, B, and C within $\mathfrak{U}$.

Represent each of the following graphically and list the elements of the set. Use A, B, and C as given above.

12. $A \cap C \cap B'$
13. $A' \cap C \cap B$
14. $A' \cap B' \cap C$
15. $A' \cap C' \cap B$
16. $A' \cap B' \cap C'$
17. Name area 3 in the figure on pg. 48.

18. Name area 4 in the figure on pg. 48
19. Name area 6 in the figure on pg. 48
20. Name area 7 in the figure on pg. 48

Answers

2.

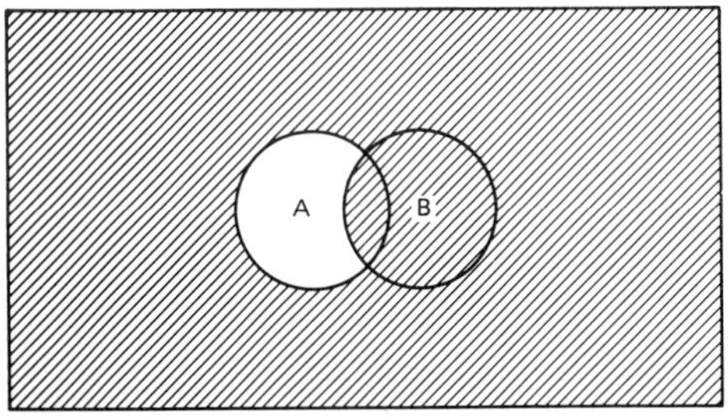

4.

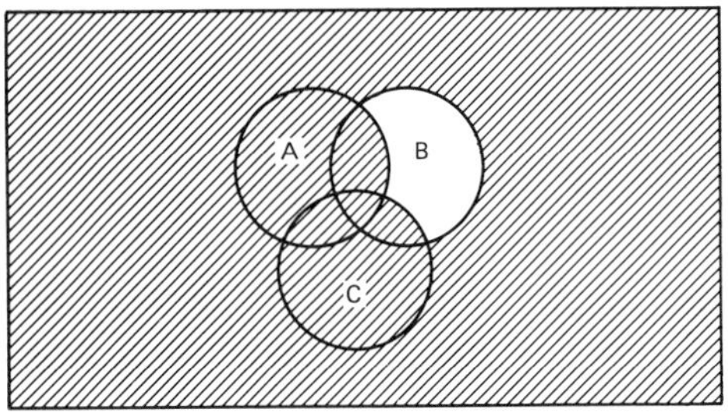

6.

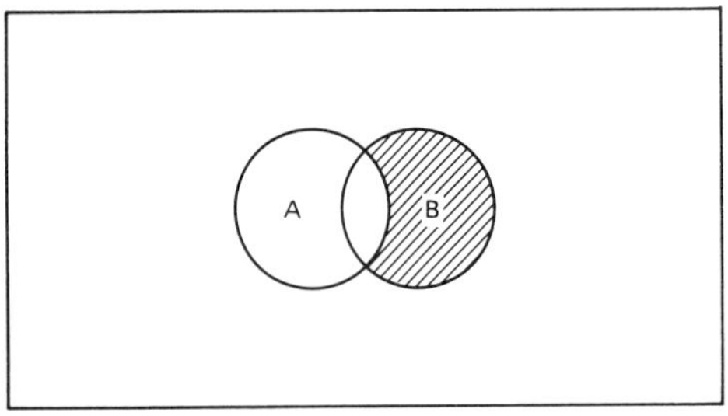

8.

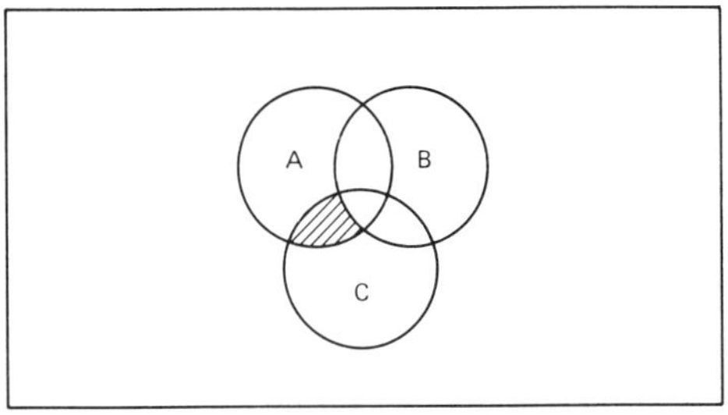

10.

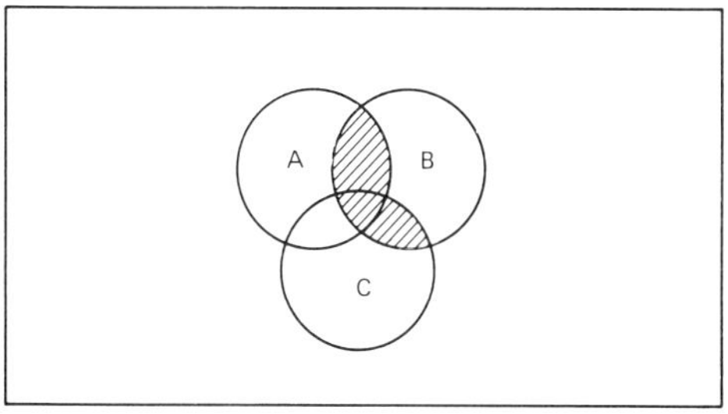

12.

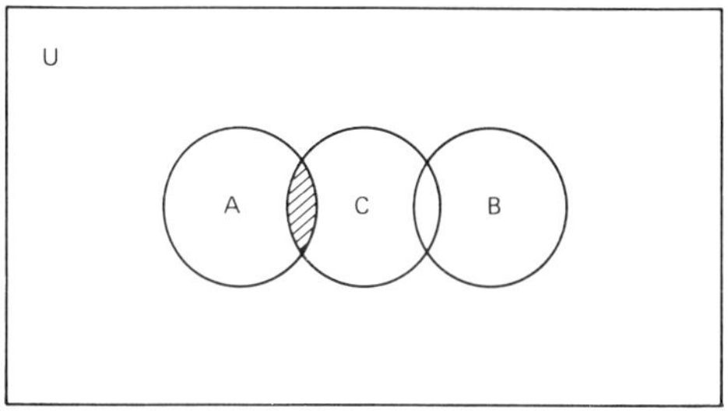

$\{2, 4\}$

14.

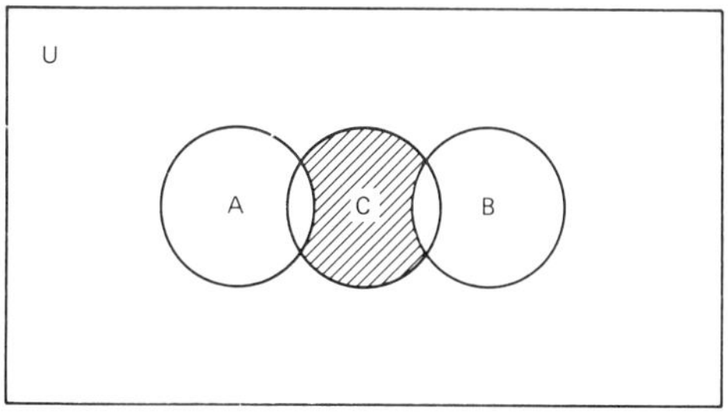

$\{0\}$

16.

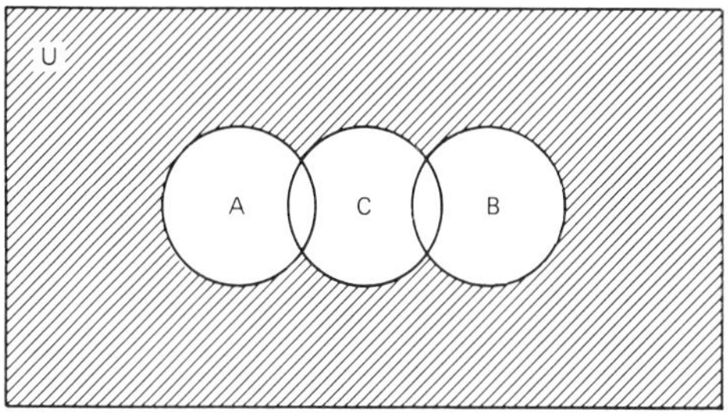

$\{\ \}$

18. $A \cap B \cap C'$

20. $A \cap B' \cap C'$

2.4 DECISIONS USING VENN DIAGRAMS

Venn diagrams clarify the elements involved in a set description and determine the membership of a set. Venn diagrams can also demonstrate the validity of an argument; however they never constitute a proof for validity.

If an answer to whether a statement is valid is *maybe*, logic classifies the statement as invalid. For a statement to be valid, it must be true in all situations.

All x are y is shown in a Venn diagram as two sets; the set containing x is a subset of y. The Venn diagram for this statement is:

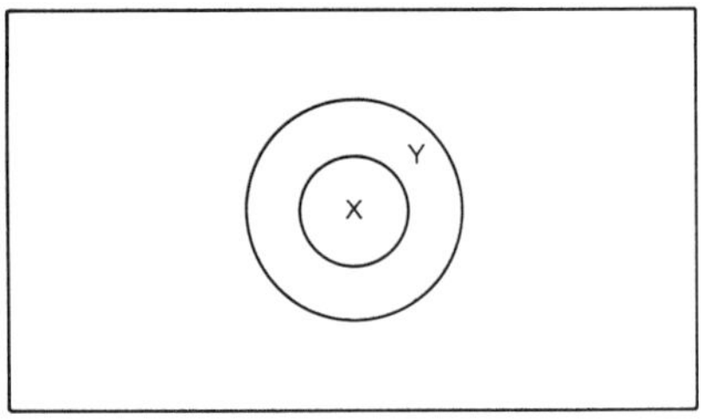

No x are y is shown on a Venn diagram as two disjoint sets.

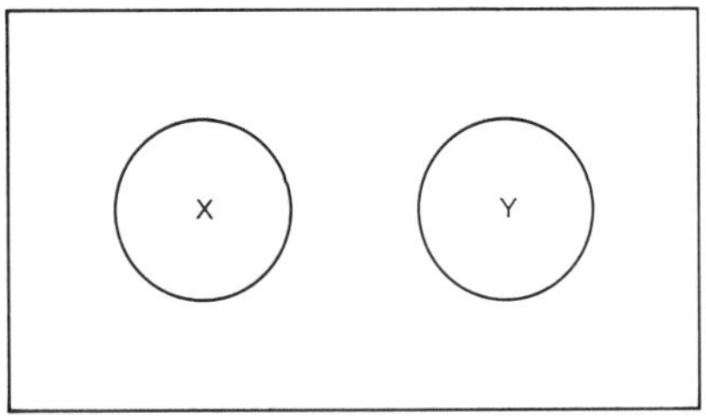

Some x are y is shown on a Venn diagram as two intersecting sets: the intersection is not empty.

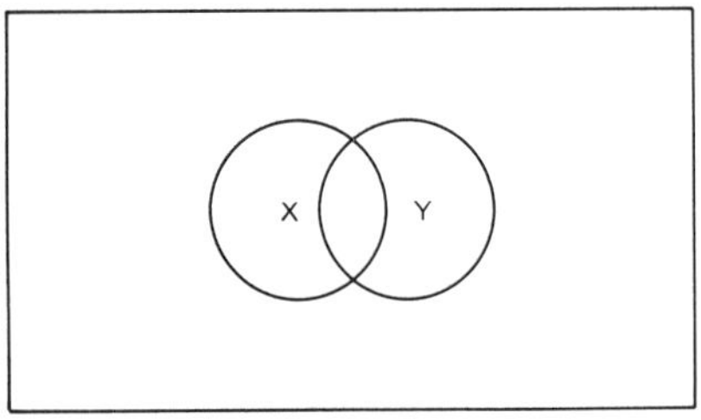

It is important not to add conditions to a statement. The following is an example illustrating this fact.

Given the two valid statements:

All bachelors are rich.
Some bachelors do not drink.

To draw a Venn diagram to represent these two statements, three sets will be used. B represents bachelors, R represents rich people, and D represents drinking people. The Venn diagrams can be drawn in two ways; the statements do not tell us which is correct so both must be considered.

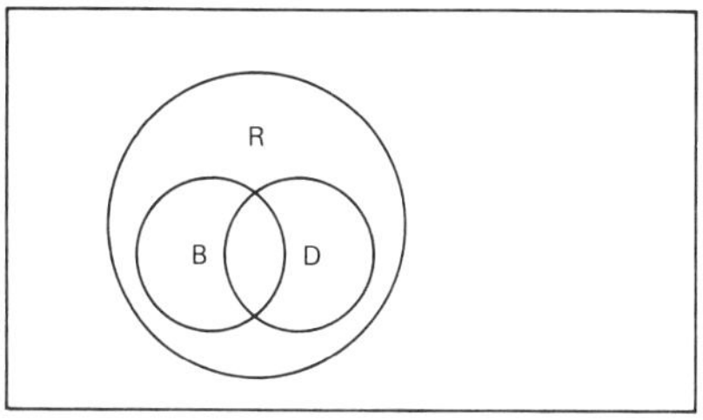

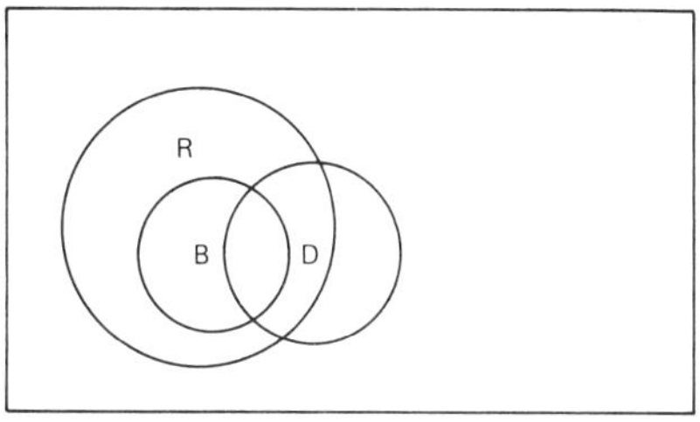

If the problem is to determine the validity of:

Some rich men do not drink.

The statement is true in both cases: Elements exist outside of D but in R. The statement is valid. If the problem is to determine the validity of:

Some men who drink are not rich.

The statement is invalid because all of D is within R in one diagram,

Example

Given:

All data processing students are ambitious. All ambitious people are rich.

Determine the validity of:

All data processing students are rich.

Draw a Venn diagram where S is the set of data processing students. A is the set of ambitious people. R is the set of rich people. Arrange the sets as described in the given

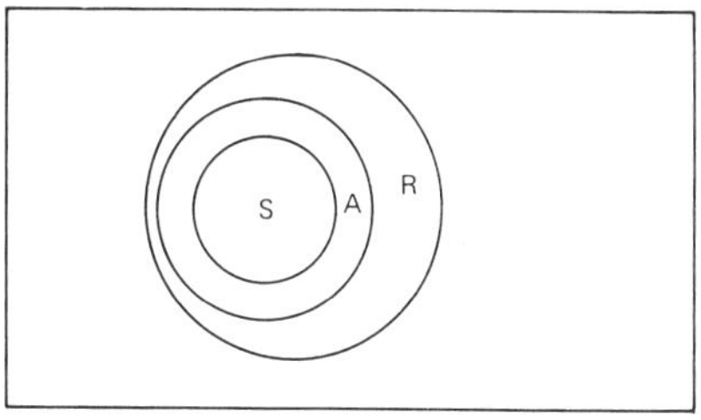

S is a subset of R, therefore it is valid to say:

All data processing students are rich.

Example

Given: No teacher is ignorant. All ignorant people are vain.
Determine the validity of:

No teacher is vain.

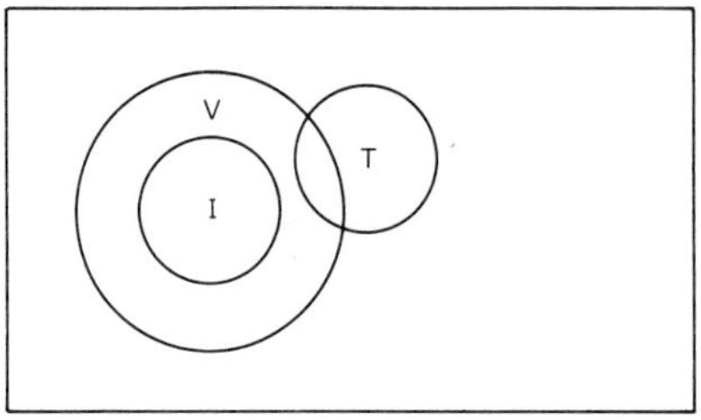

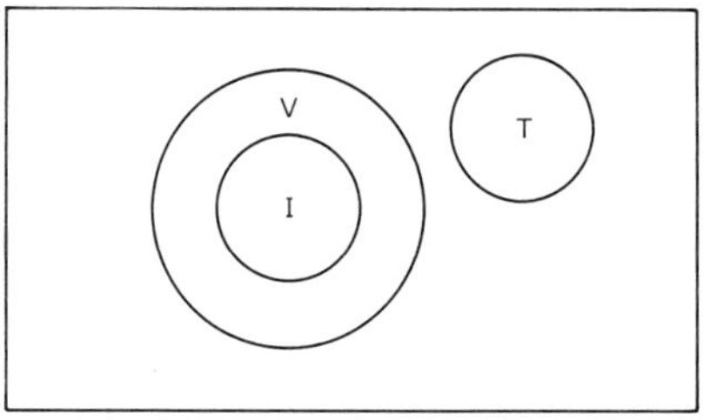

The given information does not determine the relationship between V and T. Both cases must be considered.

The statement "no teacher is vain" is invalid.

Example

Given: All people with income greater than $25,000 pay income tax. No blind person pays income tax.
Determine the validity of: No blind person has an income greater than $25,000.

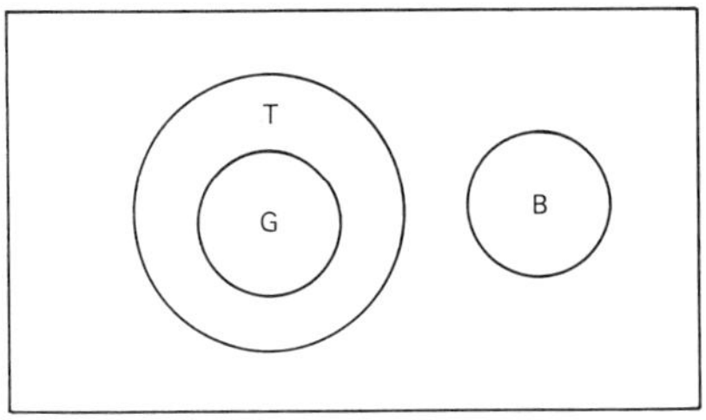

The Venn diagram demonstrates the validity of the statement. Advertising suggests invalid statements in commercials. Beautiful women smoke cigarettes. You smoke. Therefore, you must be beautiful!

The most troublesome word in determining validity is the word "some." Whenever "some" is used, check all possible arrangements for a diagram.

EXERCISES

Determine the validity of each of the following using Venn diagrams to demonstrate.

1. All books are useful. Some literature is useful.
   Test the validity of: Some literature is a book.
2. All dancers are eccentric people. All eccentric people are rich. All eccentric people are singers.
   Test the validity of: All singers are rich.
3. All people who like money love KRINKLES.
   Test the validity of: People who do not like money do not like KRINKLES.
4. All people with arthritis take aspirin. You take aspirin.
   Test the validity of: You have arthritis.
5. All good-looking people drive Cadillacs. You drive a Cadillac.
   Test the validity of: You are good looking.
6. No woman voted in the election. All union members voted in the election.
   Test the validity of: No woman is a union member.
7. All New Yorkers are Americans. Some New Yorkers are Yankee fans.
   Test the validity of: Some Americans are Yankee fans.
8. No Marine has long hair. All long-haired individuals are hippies.
   Test the validity of: No Marine is a hippie.
9. All mathematics is difficult material. Some mathematics is useful material.
   Test the validity of: Some difficult material is useful.
10. Some beautiful women are blondes. All blondes have more fun.
    Test the validity of: Some beautiful women have more fun.

Draw a Venn diagram to help decide if each of the following is true or false.

11. If $x \in (A \cap B)$, then $x \in A$.
12. If $x \in (A \cap B)$, then $x \in B$.
13. If $B \subset A$ and $x \in A$, then $x \in B$.
14. If $B \subset A$ and $x \in B'$, then $x \in A'$.
15. If $A \cap B \neq \phi$ and $A \subset C$, then $B \cap C \neq \phi$.
16. If $B \subset A$, then $A' \subset B'$.
17. If $A \subset B$, then $A' \subset B'$.
18. If $A \cap B = \phi$ and $x \in B$, then $x \in A'$.
19. If $A \cap B = \phi$ and $x \in A$, then $x \in B'$.
20. The complement of $A \cap B$ is the union of the complements of A and B.

Answers

2. Not valid
4. Not valid

6. Valid
8. Not valid
10. Valid
12. True
14. False
16. True
18. True
20. True

2.5 TRANSLATING VENN DIAGRAMS TO SETS

Often approaching material from another direction will clarify the original material as well as expand on its usefulness.

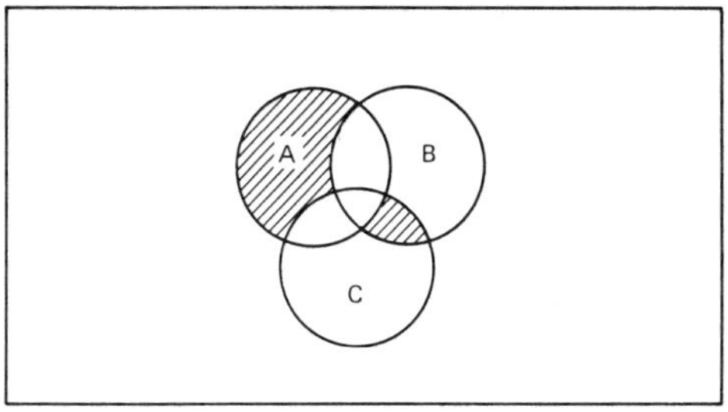

In naming the shaded area in the diagram above, take one area at a time. The upper left area is in A, not in B, and not in C, so the set notation is $A \cap B' \cap C'$. The area on the right is in B, in C, but (and) not in A, so the set notation is $B \cap C \cap A'$. The union of these two sets is

$$(A \cap B' \cap C') \cup (B \cap C \cap A')$$

The verbal translation depends on the description of subsets A, B, and C. If A represents Republican voters, B represents southerners, and C represents voters under 50 years of age, the verbal description is nonsouthern Republican voters who are over 50 years of age or southern voters under 50 years of age who are not Republicans.

If the subsets are A = football fans, B = baseball fans, and C = basketball fans, the shaded areas represent football fans who are not basketball fans and are not baseball fans or fans of basketball and baseball but not football. Once the set notation is written, the verbal description can be made according to the set description.

Example

If an entire circle is shaded, the area is named by the subset name. The set description of the Venn diagram is A.

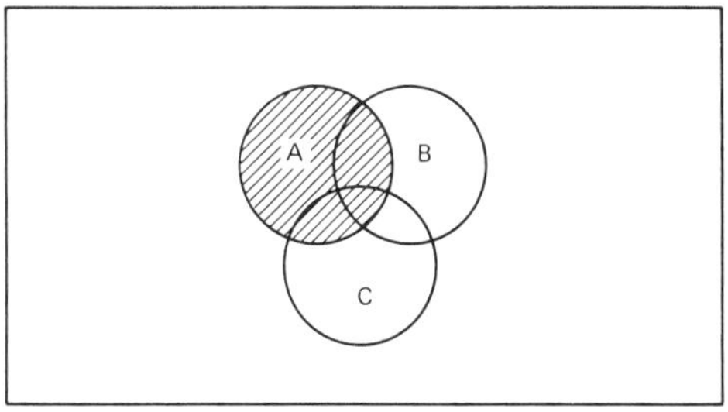

Example

If all except a region is shaded, name the area not shaded and take the complement of that. In this Venn diagram, the set description is $(A \cup B)'$.

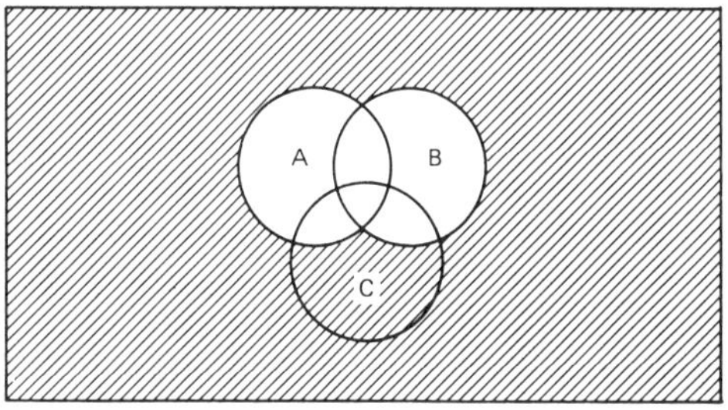

Recall that listing elements more than once does not alter the subset. Therefore, an area can be included in more than one subset in a Venn diagram without altering the overall membership.

Example

The area represented in the Venn diagram below is $(A \cap C) \cup (B \cap C)$. This set notation includes the middle area $(A \cap B \cap C)$ in each of the subsets $A \cap C$ and $B \cap C$, but the overall membership is not altered by including this area twice.

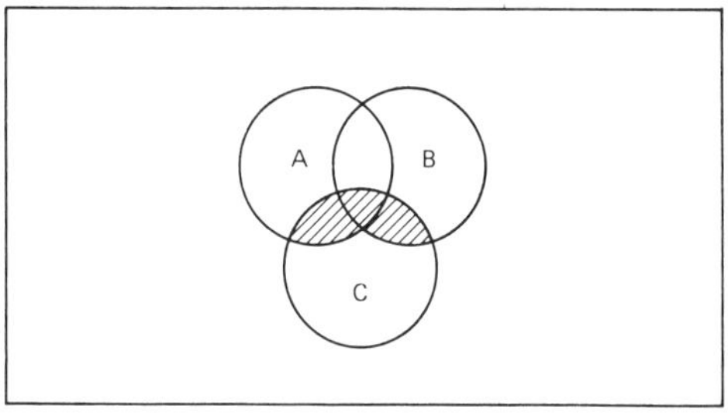

Often set notation can be simplified. However this chapter does not include simplification techniques, a later chapter introduces techniques that lend themselves to simplification of set notation. Until then, accuracy in the use of set notation is of prime importance.

EXERCISES

Name each of the following in set notation.

1. 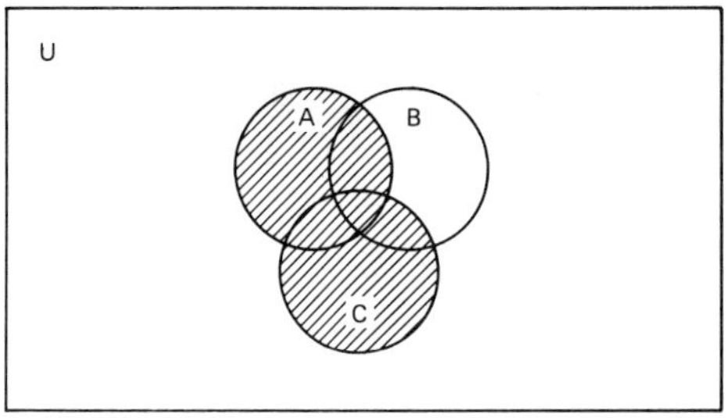

2. 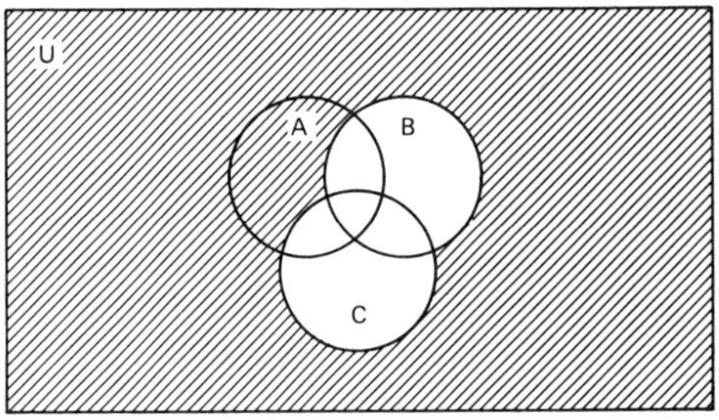

3. 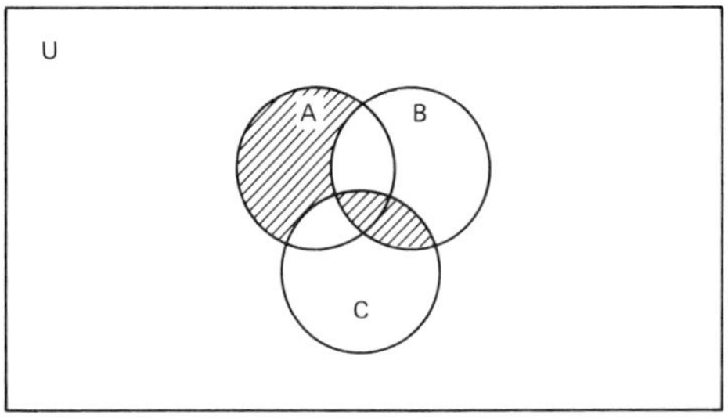

4.

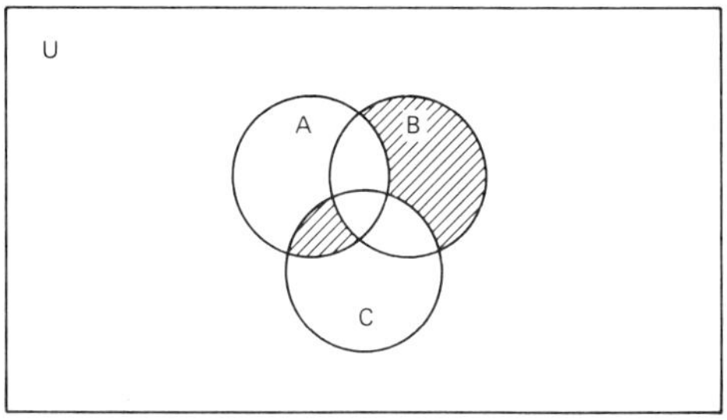

5.

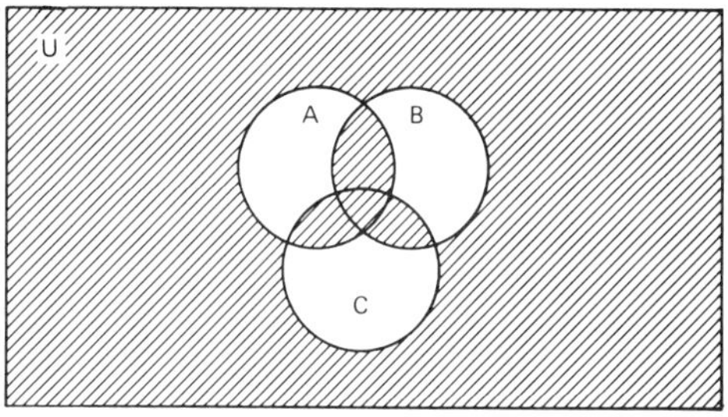

6.

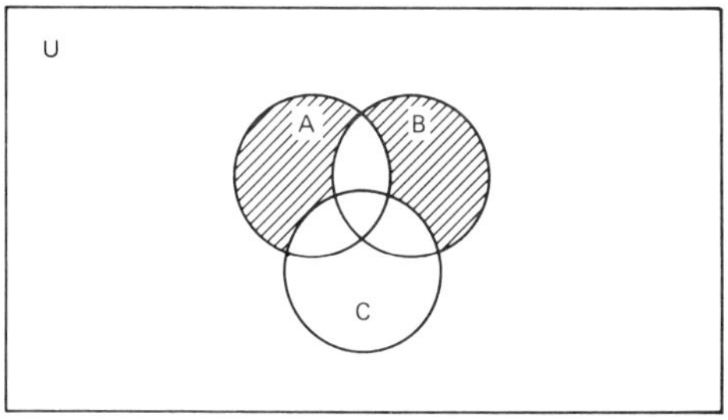

7.

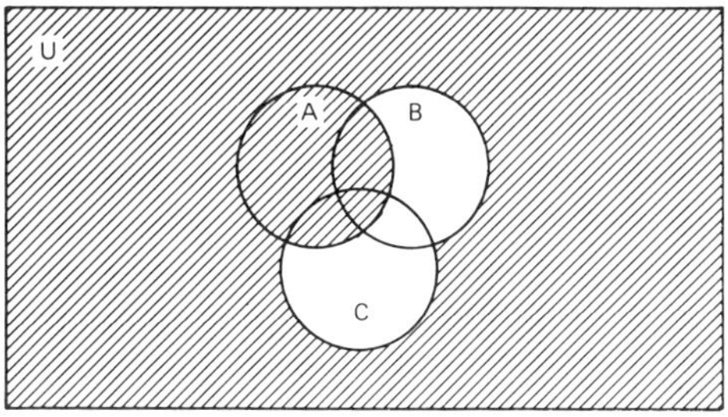

8.

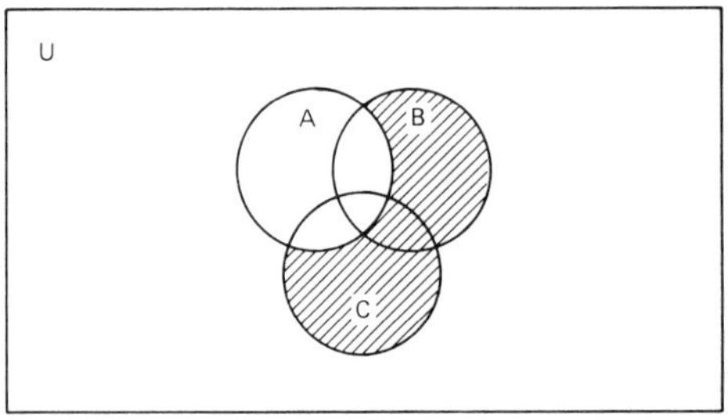

9.

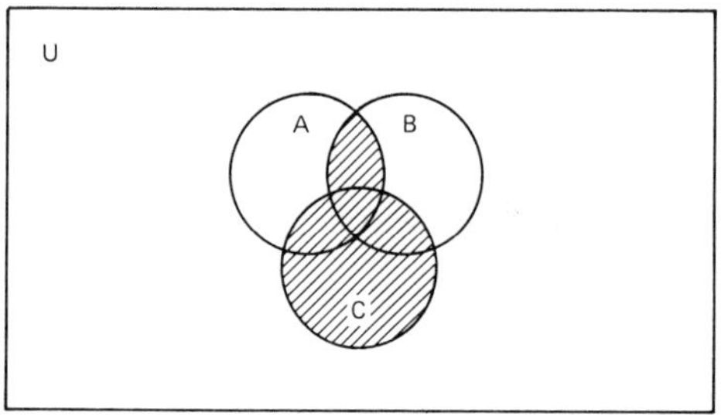

10.

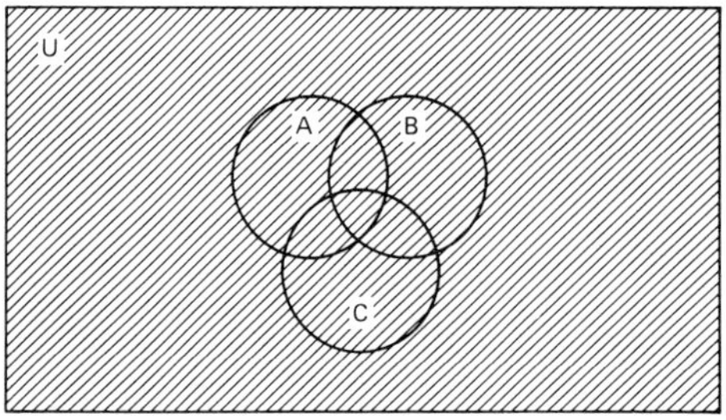

11–20. Give a verbal description for each of the shaded areas given in Exercise 1–10 for the following.

$\mathfrak{U}$ is the set of students at ABC University.

A is the set of students enrolled in English.

B is the set of students enrolled in mathematics.

C is the set of students enrolled in computer programming.

Answers

2. $(B \cup C)'$
4. $(A \cap C \cap B') \cup (B \cap A' \cap C')$
6. $(A \cap B' \cap C') \cup (B \cap A' \cap C')$

8. $(B \cap A') \cup (C \cap A')$
10. $\mathcal{U}$
12. All students not taking math or computer programming.
14. Students taking English, computer programming, but not mathematics or math students not taking English or computer programming.
16. Students enrolled in English but not in math or computer programming or students enrolled only in math.
18. Students enrolled in math but not English or students in computer programming but not English.
20. All students at ABC University.

2.6 NUMBER OF ELEMENTS IN A FINITE SET

Venn diagrams are useful in determining the number of elements in a set or subset. The population of a subset can be determined from a formula or by applying the available numbers to a Venn diagram and allowing the picture to determine the populations required. If the required formulas are listed, a diagram can be used to demonstrate the formulas, and the problems can then be solved by using the formulas and Venn diagrams. The population of A is denoted by $n(A)$. The formula for the population of the union of two subsets is $n(A \cup B) = n(A) + n(B) - n(A \cap B)$. Also, the formula for the population of the union of three subsets is $n(D \cup E \cup F) = n(D) + n(E) + n(F) - n(D \cap E) - n(D \cap F) - n(E \cap F) + n(D \cap E \cap F)$.

Draw a two-subset Venn diagram and label the areas I, II, III, and IV. Draw a three-subset Venn diagram and label the areas 1–8 as shown in section 2.3.

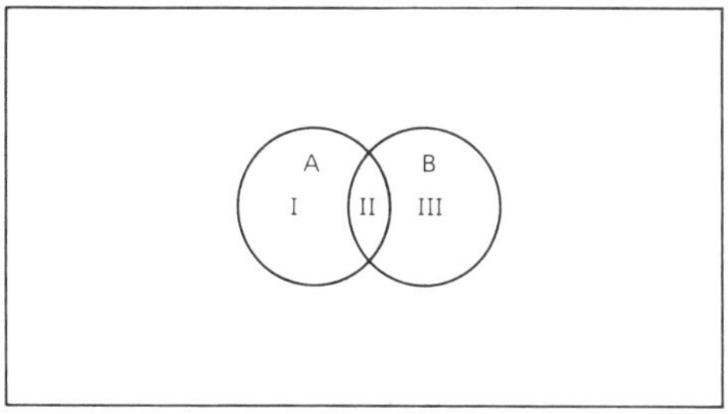

Diagram A

$$n(A) = \text{number in I and II} = n(\text{I}) + n(\text{II})$$

$$n(B) = \text{number } n(\text{II}) \text{ and } n(\text{III})$$

$$n(A \cup B) = \text{number in } n(\text{I}) + n(\text{II}) + n(\text{III})$$

$$n(A \cap B) = \text{number II}$$

Substituting in the formula for two subsets:

$$n(\text{I}) + n(\text{II}) + n(\text{III}) = n(\text{I}) + n(\text{II}) + n(\text{II}) + n(\text{III}) - n(\text{II})$$

$$n(\text{I}) + n(\text{II}) + n(\text{III}) = n(\text{I}) + n(\text{II}) + n(\text{III})$$

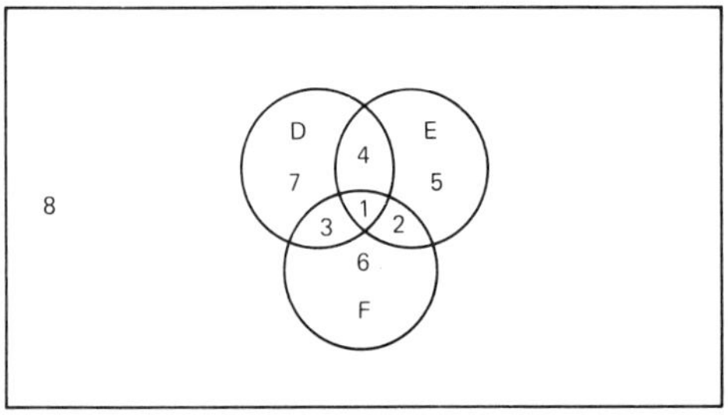

Diagram B

A like substitution can be done for the second diagram with the areas representing populations labeled 1-8:

$$n(D) = n(1) + n(3) + n(4) + n(7)$$

$$n(E) = n(1) + n(2) + n(4) + n(5)$$

$$n(F) = n(1) + n(2) + n(3) + n(6)$$

$$n(D \cap E) = n(1) + n(4)$$

$$n(D \cap F) = n(1) + n(3)$$

$$n(E \cap F) = n(1) + n(2)$$

$$n(D \cap E \cap F) = n(1)$$

Substituting these numbers in the formula for three sets produces:

$$n(1) + n(2) + n(3) + n(4) + n(5) + n(6) + n(7)$$

$$= n(1) + n(3) + n(4) + n(7) + n(1) + n(2)$$

$$+ n(4) + n(5) + n(1) + n(2) + n(3) + n(6)$$

$$- n(1) - (n)(4) - n(1) - n(2) - n(1) - n(3) + n(1)$$

$$n(1) + n(2) \ldots n(7) = n(1) + n(2) + \ldots n(7)$$

An example best illustrates the use of formulas and Venn diagrams.

Example

In a two-party election district where 135 people voted, 67 voted for at least one Republican and 84 people voted for at least one Democrat. How many voted for candidates from both parties?

$\mathcal{U}$ consists of all voters; $n(\mathcal{U}) = 135$.

R represents those who voted for at least one Republican, $n(R) = 67$.

D represents those who voted for at least one Democrat; $n(D) = 84$. Also $n(R \cup D) = 135$

The formula states $n(R \cup D) = n(R) + n(D) - n(R \cap D)$. Substituting the known values gives $135 = 67 + 84 - n(R \cap D)$. Solving this statement for $n(R \cap D)$ gives a population of 16 for the subset representing those who voted for a Democrat candidate and a Republican candidate.

The information is shown on the given Venn diagram. Because 67 + 84 exceeds the universal population of 135, it can be determined the intersection must have 16; the number voting exclusively Republican is $67 - 16 = 51$ and the number voting exclusively Democratic is $84 - 16 = 68$. The subset of $(R \cup D)'$ is empty because all people being considered voted either Republican or Democratic. Many election statistics come from analysis of data such as this.

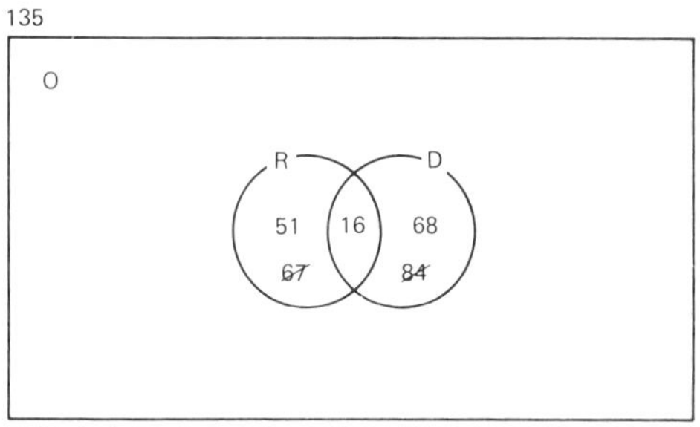

If the example involves percent, the universal set must be 100%.

Example

A government committee reported that, among the students using marijuana, LSD, or heroin at a certain university, 90% used marijuana, 6% used LSD, 7% used heroin, 4% used marijuana and LSD, 5% used marijuana and heroin, 2% used heroin and LSD, and 1% used all three. Do you find the committee figures consistent?

Venn Diagram

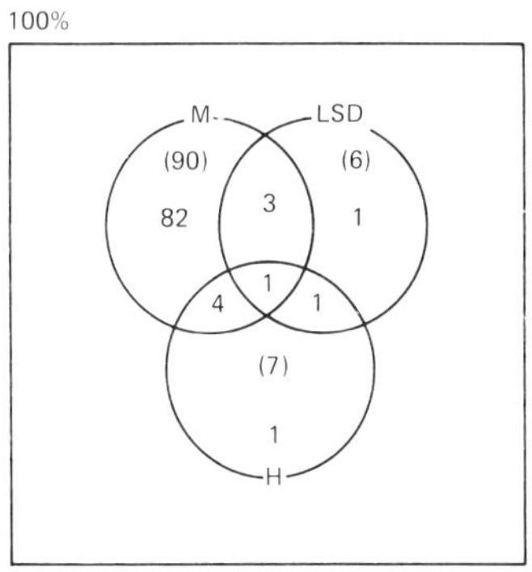

( ) = # before intersections subtracted.

82 + 3 + 1 + 4 + 1 + 1 + 1 = 93%

Population of Sets

$n(\mathfrak{U}) = 100\%$ $\quad n(M \cap L \cap H) = 1\%$

$n(L \cap M) = 4\%$ $\quad n(M \cap H) = 5\%$

$n(L \cap H) = 2\%$ $\quad n(M) = 90\%$

$n(L) = 6\%$ $\quad n(H) = 7\%$

$$n(M \cup L \cup H) = n(H) + n(L) + n(M) - n(M \cap H) - n(L \cap M) - n(L \cap H) + n(M \cap H \cap L)$$

$$7\% + 6\% + 90\% - 4\% - 2\% - 5\% + 1\% = 93\%$$

The figures are not consistent because the total population must equal 100%. The Venn diagram represents the population percents.

Venn diagrams can often show additional information that the formulas do not show. In this example, it is possible to find that 82% use only marijuana, 1% use only LSD, and 1% use only heroin.

General technique for using Venn diagrams

1. Work from the intersection population outward.
2. When the number in a subset is known, subtract the population of any intersections within the subset to obtain the population in a subset and not in any other subset.
3. Total the populations of the areas (including the area outside the subset when appropriate) to give the population of the universal set.

EXERCISES

Solve both by formula and Venn diagram.

1. There are 100 students in a particular school. There are 11 students enrolled in math, history, and English; 18 students are taking English and math; 23 students are taking English and history; and 14 students are enrolled in math and history. A total of 40 students are enrolled in math, 39 students are taking history, and 25 students are not enrolled in any of these. How many students are taking only English?
2. In a town of 500 taxpayers, 282 pay income tax, 421 pay property tax, and 23 are over 65 and exempt from paying tax. How many taxpayers pay both income and property tax?
3. In a telephone survey of 100 people, 82 answered yes to the question: Do you drive a foreign-made car?; 64 answered yes to the question: Do you drive an American-made car?; and 3 answered no to the question: Do you drive a car? How many of those surveyed drove both a foreign-made and an American-made car?
4. In a recent survey it was found that 12% of those surveyed read a weekly news magazine, 72% watch television news, 63% read newspapers, 7% read a newspaper and read a weekly news magazine, 40% watch television news and read a newspaper, 8% watch television news and read a weekly news magazine, and 5% watch television news, read a newspaper, and read a weekly news magazine, and 3% do not bother with reading or listening to any news. What percent of those surveyed read only the newspaper?
5. In an advertising survey of a hundred people, 70 people were found to take aspirin and 30 people were found to use both aspirin and bufferin. All people surveyed used either aspirin or bufferin. Can the advertisement honestly state that, more people use aspirin than Bufferin?
6. It is known that the percents of football fans and baseball fans are equal and 28% like both football and baseball. What percent are football fans?

7. One hundred data processing students are enrolled in either MAT 114 or DAT 112. It is known that 48 students are taking MAT 114 and 78 students are taking DAT 112. How many are enrolled in both courses.
8. You have been engaged to conduct a telephone survey and will be paid 50¢ for each call made. The results you turn in are as follows:

   150 households have members that smoke.
   355 households have members that drink beer.
   475 households have members that watch television.
   106 households do not participate in any of the above.
   116 households have members that drink beer and smoke.
   299 households have members that drink beer and watch television.
   85 households have members that watch television and smoke.
   60 households have members that watch television, smoke, and drink beer.

   How much were you paid for this telephone survey?

Answers

2. 226
4. 21%
6. 64%
8. $323

CHAPTER 2 REVIEW EXERCISES

Given:

$$\mathcal{U} = \{1, 2, 3, 4, 5, 6, 7, 8, 9\}$$

$$A = \{1, 2, 3, 4, 5, 6\} \qquad B = \{2, 4, 6\} \qquad C = \{1, 5, 6, 7\}$$

Classify each of the following as *true* or *false*. If false, change the statement to make it true.

1. $A \cup C = A$
2. $B \cap C = \emptyset$
3. $B \cap C \subset A$
4. $C' \cup A = \mathcal{U}$
5. $B' \cap C' = (B \cup C)'$
6. $7 \in (B \cup C)$
7. $4 \in B, 4 \in A, 4 \in A \cap B$
8. $C \cap C' = \emptyset$
9. $A \cap B \cap C = \emptyset$
10. Draw a Venn diagram representing the $\mathcal{U}, A, B$, and C above and include the elements in each area of the diagram.

Use the given Venn diagram. Classify as *true* or *false*.

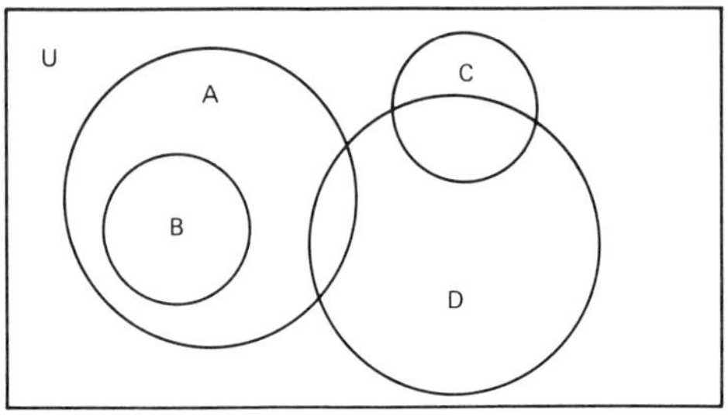

11. All A are B.
12. Some D are B.
13. No C are B.
14. No D are A.
15. $B \subset A$

Use the following verbal descriptions to translate each of the following into set notation.

$\mathfrak{U}$ is the set of four-legged animals

A is the set of turtles

B is the set of orange-colored animals

C is the set of female animals

16. Four-legged male animals that are not turtles
17. Turtles or female orange-colored animals
18. Female animals or orange-colored animals.
19. Orange female turtles
20. Male four-legged animals that are not turtles or orange-colored animals

Write the set notation for each of the Venn diagrams.

21.

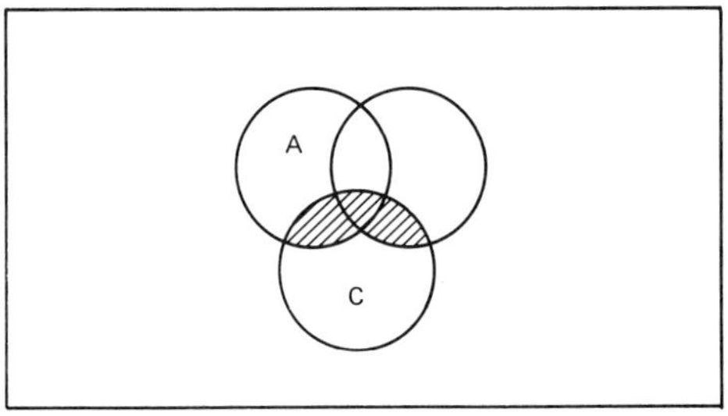

22.

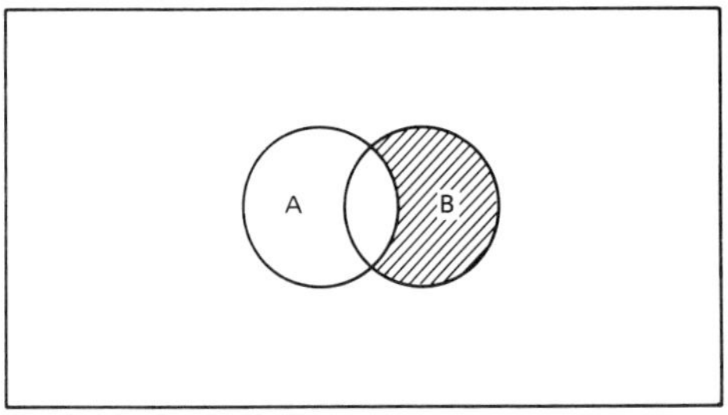

23.

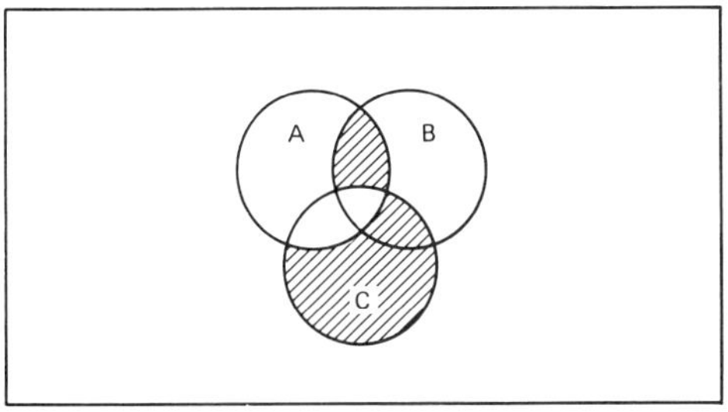

Draw a Venn diagram for each.

24. $A' \cup B$

25. $A \cap B'$

26. $A' \cap (B \cup C')$

Determine the validity of each conclusion.

27. All cats are pets. Some pets are rabid.
    Cats are not rabid.

28. Some senators are wealthy. All wealthy people pay taxes.
    All senators pay taxes.

29. All thunderstorms bring rain. Some hurricanes produce thunderstorms.
    Some hurricanes produce rain.

Determine the number of elements in the following.

30. Of 100 dentists polled, 75 recommended brand X toothpaste, 60 recommended orange juice daily, and 10 made no recommendations. How many recommended both brand X toothpaste and orange juice?

Chapter 2 Review Answers

2. False, $B \cap C = \{6\}$
4. False, $C' \cup A = \{1, 2, 3, 4, 5, 6, 8, 9\}$
6. True
8. True

10.

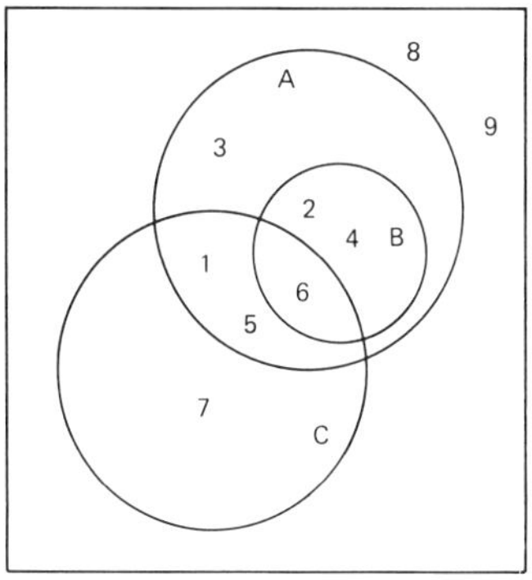

12. False
14. False
16. $C' \cap A'$
18. $C \cup B$
20. $(C' \cap A') \cup B$
22. $B \cap A'$
24.

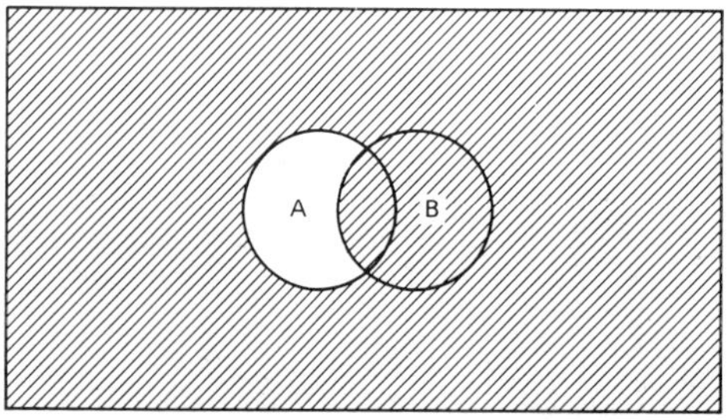

26.

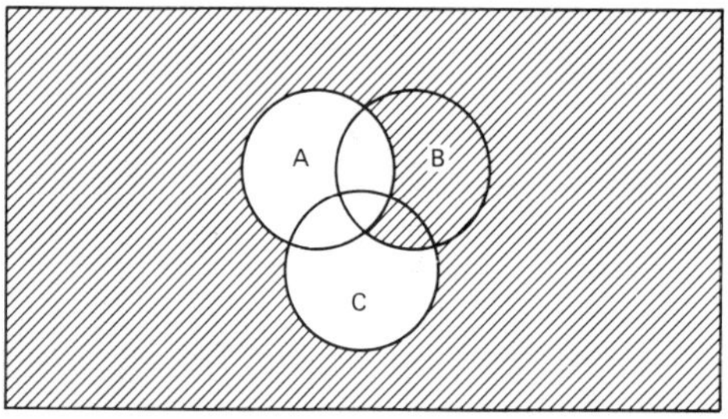

28. Not valid
30. 45

Logic

CHAPTER 3

3.0 INTRODUCTION

Logic is the science of correct reasoning, a science that has been pursued since the birth of civilization. Symbolic logic has been developed over a period of years and used as a tool in analysis and problem solving. The first mathematician credited with symbolic logic is Gottfried Leibniz. George Boole furthered the study started by Leibniz. Alfred Whitehead and Bertrand Russell popularized logic in the twentieth century. The computer age has emphasized the need for correct reasoning; and symbolic logic is a method of developing this science.

The purpose of symbolic logic is to set up a notation that produces valid conclusions from true premises. A set of rules is determined. New truths are deduced from existing truths. Logic demands precision in definitions and consistency in using the definitions. Applications of logic can point to contradictions, equivalences, and necessary conditions. A study of logic can aid in communicating clearly and precisely with others. Logic is used as a guide in judging the validity of statements or the reasoning of others.

The chapter is designed to:

1. Provide a study of the basic connectives in logic and the rules that govern each connective.
2. Provide the mechanics of constructing and interpreting truth tables.

3. Provide the tools for analyzing verbal statements and translating them into valid and efficient flowcharts.
4. Promote good reasoning habits and sharpen problem-solving skills.

3.1 BASIC INGREDIENTS OF LOGIC

The basic ingredient of logic is a statement.

By definition a *statement* is a sentence that can be classified as true or false. Notice that a statement is more than a sentence. Some examples of sentences are:

1. The moon is made of green cheese.	A statement.
2. $7 + x = 8$.	Not a statement; it is neither true nor false until x is identified.
3. Shut the door!	Not a statement; it cannot be classified as true or false.
4. Today is Wednesday.	A statement.
5. What time is it?	Not a statement; questions are not classified as statements.
6. $3 + 2 = 5$	A statement.

In symbolic logic, verbal statements are symbolized by small letters; tradition suggests the middle letters of the alphabet ($p, q, \ldots$).

Example

The statement "I have a penny" is a simple statement. It can be symbolized by p, where p represents "I have a penny." Thus p can have a value of true or false. In a diagram, this is shown as:

p
T
F

The statement "I have a quarter" can be symbolized as q.

Compound statement. A *compound statement* is a statement involving one or more simple statements and one or more connectives. This section considers four connectives—AND, OR, exclusive OR, and NOT.

AND. *AND* is known in English as a *conjunctive.* The logic symbol is $\wedge$. The AND connects two simple statements. In order for such a compound statement to be true, each of the simple statements must be true. The diagram that defines the AND connective is as follows:

p	q	$p \wedge q$
T	T	T
T	F	F
F	T	F
F	F	F

This diagram gives all of the possible combinations of true and false values available for the simple statements p and q. The truth value of the compound sentence $p \wedge q$ is defined in each situation.

Example

I have a penny and a quarter. This is a compound statement made up of the statement "I have a penny" and the statement "I have a quarter." These two statements are connected with an AND. Using p and q as defined previously, the compound statement is $p \wedge q$. The definition of AND indicates this statement is true only when "I have a penny" is true and "I have a quarter" is also true.

The definition of the connective AND is consistent with the English composition use of the word *and*.

OR. *OR* is the *disjunctive* and is symbolized by $\vee$. The English use of the word *or* leads to ambiguity when applied to logic. In logic the meaning of the statements connected by OR has no bearing on the overall validity of a compound statement. An example is probably the best way to show the ambiguity.

Example

I have a dog or cat. This statement is true if "I have a dog" is true and "I have a cat" is false; if "I have a dog" is true and "I have a cat" is true; or if "I have a cat" is true and "I have a dog" is false.

Example

I am an American or a Russian. This is a true statement if "I am an American" is true and "I am a Russian" is false or if "I am a Russian" is true and "I am an American" is false. It is impossible to be both American and Russian.

In logic, a connective is needed for situations where interpretation of the content suggests inconsistency. To designate the difference in the two ORs, two connectives are available in logic.

The OR symbolized by $\vee$ is known as *inclusive or*. It allows a statement to be true if one or both of the simple statements connected by OR are true.

EOR. *EOR* is known as *exclusive or*, symbolized by $\underline{\vee}$. It allows a compound OR statement to be true when only one of the simple statements is true. To designate this, the word *either* will be used in connection with *or* or *but not both* will follow the statement.

The truth values for the inclusive and exclusive OR are:

p	q	$p \veebar q$	$p \vee q$
T	T	F	T
T	F	T	T
F	T	T	T
F	F	F	F

NOT is known as a negation and is symbolized in logic by a ~ preceding the statement, a line over the statement affected by the NOT, or a prime (′) following the letter: NOT $p = {\sim}p = \bar{p} = p'$. This book uses the ~ to denote NOT.

In English, the sentence "Today is not Tuesday" is a simple statement. In logic, the affirmative statement "Today is Tuesday" is the simple statement, r. "Today is not Tuesday" is symbolized in logic by NOT r or ${\sim}r$.

NOT reverses the truth value of a statement:

p	${\sim}p$
T	F
F	T

A compound statement preceded by a NOT is translated as "it is not the case that" or "it is not true that." These phrases indicate that the truth value of the compound statement is reversed.

Example

${\sim}(p \vee q)$

p represents "I have a penny"

q represents "I have a quarter"

${\sim}(p \vee q)$ represents "It is not true that I have a penny or a quarter."

Caution must be exercised when translating a compound statement preceded by NOT. The distributive property does not hold for NOT: ${\sim}(p \vee q)$ and ${\sim}p \vee {\sim}q$ have different meanings AND different truth values.

EXERCISE

Which of the following are statements?

1. The hill is flat.
2. Are you hungry?

3. The older you get, the wiser you become.
4. The course is useful to potential buyers.
5. Go west, young man, go west!
6. John Doe was our tenth president.
7. Do you have the time?
8. Shake a leg.
9. Set notation includes complements.
10. Binary numbers have the digits 0 and 1.

Let p represent "Polly is young" and q represent "David is sleepy." Assume old is not young and alert is not sleepy. Symbolize each of the following:

11. Polly is old and David is sleepy.
12. Polly is young or David is alert.
13. It is not the case that Polly is young and David is sleepy.
14. Either Polly is young or David is alert.
15. Polly is old or David is alert, but not both.

Let p represent "The employee is over 50" and q represent "The employee is female." Symbolize each of the following:

16. The employee is not over 50 and is male.
17. The employee is over 50 or male.
18. The employee is male or over 50, but not both.
19. Either the employee is over 50 or female.
20. It is not true that the employee is a male over 50.

Answers

2. Not a statement
4. A statement
6. A statement
8. Not a statement
10. A statement
12. $p \vee \sim q$
14. $p \underline{\vee} \sim q$
16. $\sim p \wedge \sim q$
18. $\sim q \underline{\vee} p$
20. $\sim(\sim q \wedge p)$

3.2 TRUTH TABLES

A truth table is a device that generates the truth values of a compound statement by using the definitions of the connectives and all possible combinations of the truth values of the simple statements. It is imperative to know the truth values of all the fundamental connectives. The following table gives a summary of all of the definitions of the connectives:

p	q	$p \wedge q$	$p \vee q$	$p \underline{\vee} q$	$\sim p$	$\sim q$
T	T	T	T	F	F	F
T	F	F	T	T	F	T
F	T	F	T	T	T	F
F	F	F	F	F	T	T

Steps for Constructing a Truth Table

1. Determine the number of simple statements. The number of lines in the truth table is 2^n, where n is the number of simple statements.
2. List the symbols for the simple statements alphabetically, one column for each letter. Start with the rightmost symbol. Fill in the column with alternating T's and F's for the determined number of lines (T, F, T, F . . .). The next column to the left has alternating TT and FF. The third column to the left has alternating $TTTT$ and $FFFF$, and so on. This method insures all possible truth-value combinations and allows for uniformity in construction. Results can be readily checked and compared.
3. Make a column for each connective, listing connectives within parentheses first and then NOT, AND, OR. If two or more connectives are alike, work from left to right.
4. Operate on each line of the truth table using the truth value of the simple or compound sentence involved and the definition of the connective.

Example

Construct a truth table for "Today is not Wednesday and tomorrow is not Friday." Let p = "Today is Wednesday" and q = "Tomorrow is Friday." The sentence is symbolized as $\sim p \wedge \sim q$

Step 1. There are two simple statements, so there are $2^2 = 4$ lines in truth table.

p	q
T	T
T	F
F	T
F	F

Step 2. There are three connectives in the compound statement. NOT is applied first and then AND.

p	q	$\sim p$	$\sim q$	$\sim p \wedge \sim q$
T	T			
T	F			
F	T			
F	F			

Step 3. The NOT connective reverses the truth value of the simple statement to which it is attached. Therefore, the column $\sim p$ reverses the truth values, line by line, of p. The same is true for q.

		A	B	C
p	q	$\sim p$	$\sim q$	$\sim p \wedge \sim q$
1. T	T	F	F	
2. T	F	F	T	
3. F	T	T	F	
4. F	F	T	T	

Step 4. Now the AND connective is applied to the $\sim p$ and $\sim q$ columns, marked A and B. Recall the rules of the AND connective. Line 1 combines F and F, which yields a F in column C. Line 2 combines a F and T, to give a F in column C. Line 3 combines a T and F, which gives F in column C. Line 4 combines a T and T, to produce a T in column C. The completed truth table is as follows:

p	q	$\sim p$	$\sim q$	$\sim p \wedge \sim q$
T	T	F	F	F
T	F	F	T	F
F	T	T	F	F
F	F	T	T	T

Step 5. Interpretation: Today is not Wednesday and tomorrow is not Friday is a true statement only when Today is Wednesday is false and tomorrow is Friday is false.

Truth tables can be used to determine if two propositions have the same truth values. When two statements have the same truth values, one can replace the other in any logical expression.

Example

It is not true that I am not old.
Let p represent "I am old." $\sim p$ represents I am not old. $\sim(\sim p)$ = It is not true that I am not old. The truth table has one statement, so a two line truth table is

needed. It has two connectives so there will be a total of three columns. The truth value of $\sim(\sim p)$ is the same as the truth value of p. This means anytime $\sim(\sim p)$ occurs, p can replace it. This is consistent with double negatives in English!

p	$\sim p$	$\sim(\sim p)$
T	F	T
F	T	F

Example

Construct a truth table for $(p \wedge \sim q) \vee (\sim p \wedge q)$. There are two simple statements so a four line truth table is needed.

p	q	$\sim p$	$\sim q$	$p \wedge \sim q$	$\sim p \wedge q$	$(p \wedge \sim q) \vee (\sim p \wedge q)$
T	T	F	F	F	F	F
T	F	F	T	T	F	T
F	T	T	F	F	T	T
F	F	T	T	F	F	F

The columns headed p, q, $\sim p$, $\sim q$ are similiar to those in example $\sim p \wedge \sim q$. The column headed $p \wedge \sim q$ takes the column values of column p AND column $\sim q$ (the first and fourth column above).

Line by line, this involves $T \wedge F = F$, $T \wedge T = T$, $F \wedge F = F$, and $F \wedge T = F$. The entries in the next column, headed $\sim p \wedge q$, result from combining column $\sim p$ with column q using the AND connective. The last column is generated from the two columns just to the left, connecting the truth values with OR.

The final column of the truth table has values $FTTF$. This is the identical truth value for the EOR connective. Therefore, it is possible to state $(p \wedge \sim q) \vee (\sim p \wedge q) = p \veebar q$.

A *tautology* is a statement which is always true. A *contradiction* is a statement which is always false.

Example

Find the truth values for $q \vee \sim q$, and $q \wedge \sim q$.

q	$\sim q$	$q \vee \sim q$	$q \wedge \sim q$
T	F	T	F
F	T	T	F

The truth table indicates $q \vee \sim q$ has truth values consisting only of T's. The statement is always true and is classified as a tautology. A verbal example is "Today is Tuesday" or "Today is not Tuesday."

The truth table indicates $q \wedge \sim q$ has truth values consisting only of F's. The statement is always false and is classified as a contradiction. A verbal example is "Today is Tuesday and today is not Tuesday."

EXERCISES

Construct a truth table for each of the following:

1. $\sim p \wedge q$
2. $p \vee \sim q$
3. $\sim p \wedge \sim q$
4. $\sim p \vee \sim q$
5. $\sim (p \vee q)$
6. $\sim (p \wedge q)$
7. $p \wedge \sim p$
8. $p \vee \sim p$
9. $p \wedge (p \vee q)$
10. $p \vee (p \wedge q)$
11. From the results of Exercises 1-10, list any statements that are logically equal.
12. From the results of Exercises 1-10, list any tautologies or contradictions.

Let

p represent "Microcomputers are the salvation of the world."

q represent "Happiness is the key to the future."

13. Translate Exercise 1 and tell when it is true.
14. Translate Exericse 2 and tell when it is true.
15. Translate Exercise 3 and tell when it is true. Write this statement in two ways.
16. Translate Exercise 4 and tell when it is true. Write this statement in two ways.
17. Translate Exercise 9 and write in simplified form.
18. Translate Exercise 10 and write in simplified form.

Answers

2.

p	q	$\sim q$	$p \vee \sim q$
T	T	F	T
T	F	T	T
F	T	F	F
F	F	T	T

4.

p	q	$\sim p$	$\sim q$	$\sim p \vee \sim q$
T	T	F	F	F
T	F	F	T	T
F	T	T	F	T
F	F	T	T	T

6.

p	q	$p \wedge q$	$\sim(p \wedge q)$
T	T	T	F
T	F	F	T
F	T	F	T
F	F	F	T

8.

p	$\sim p$	$p \vee \sim p$
T	F	T
F	T	T

10.

p	q	$p \wedge q$	$p \vee (p \wedge q)$
T	T	T	T
T	F	F	T
F	T	F	F
F	F	F	F

12. Exercise 7 is a contradiction; Exercise 8 is a tautology.
14. "Microcomputers are the salvation of the world or happiness is not the key to the future." This statement is true at all times except when microcomputers are not the salvation of the world and happiness is the key to the future.
16. "Microcomputers are not the salvation of the world or happiness is not the key to the future." It is not true that microcomputers are the salvation of the world and happiness is the key to the future. This statement is true at all times except when microcomputers are the salvation of the world and happiness is the key to the future.
18. "Microcomputers are the salvation of the world or microcomputers are the salvation of the world and happiness is the key to the future." This simplifies to "Microcomputers are the salvation of the world."

3.3 IMPLICATION

The sentence that most affects decisions in today's world is the conditional. An example of a conditional statement is: If you pass this course, then you may continue in your curriculum. A conditional statement is recognized by the words if and then. The *if* part of the statement is the hypothesis, or condition; the *then* part of the statement is the conclusion, or consequence. An arrow is used to symbolize a conditional or an implication. The condition precedes the arrow and the arrow points to the conclusion.

p represents "You pass this course."

q represents "You may continue in your curriculum."

$p \rightarrow q$ represents "If you pass this course, then you may continue in your curriculum."

$q \rightarrow p$ represents "If you continue in your curriculum, then you pass this course."

The truth value of an implication is defined by the following table:

p	q	$p \rightarrow q$
T	T	T
T	F	F
F	T	T
F	F	T

The definition of the truth values of an implication may appear to be in conflict with the English usage of *if* and *then*. The sense of the statements within an implication helps determine the validity of a statement in English. In logic, the connective alone determines the truth value. An implication in logic does not guarantee a cause-and-effect relationship.

The easiest way to predict the truth value of an implication in logic is to check the truth value of the conclusion. If the conclusion is true, the implication is true. If the condition and conclusion are both false, the implication is also true. "If the moon is made of green cheese, then rats inhabit the moon" is a true statement. The condition is false and the conclusion is false. The definition of the truth value of an implication for these conditions guarantees the statement is true.

There are various verbal translations for $\rightarrow$. Care must be taken to identify the condition within the statement. The condition does not always appear first in a statement. For example, let p represent "The sun shines" and q represent "The moon is yellow." Some of the different ways to write these sentences as an implication and the ways to symbolize the implication are as follows:

1. "If the sun shines, the moon is yellow."
   (The word *then* is often omitted; *if* must be present.)

$$p \longrightarrow q$$

2. "The sun shines if the moon is yellow."

$$q \longrightarrow p$$

3. "The sun shines only if the moon is yellow."
   (The statement following *only if* is the conclusion.)

$$p \longrightarrow q$$

4. "For the sun to shine it is sufficient that the moon is yellow."
   (The statement following *it is sufficient* is the condition.)

$$q \longrightarrow p$$

5. "The moon is yellow when the sun shines."
   (*When* is similiar to *if*.)

$$p \longrightarrow q$$

The truth values for a conditional consist of three *T*'s and one *F*. The same is true for OR connective, as verified by a truth table.

$$p \longrightarrow q \quad \text{is equivalent to} \quad \sim p \vee q$$

p	q	$p \rightarrow q$	$\sim p$	$\sim p \vee q$
T	*T*	*T*	*F*	*T*
T	*F*	*F*	*F*	*F*
F	*T*	*T*	*T*	*T*
F	*F*	*T*	*T*	*T*

Same truth values (columns $p \rightarrow q$ and $\sim p \vee q$)

This means the statement "If it rains today, the picnic will be canceled" is logically equivalent to "It does not rain today or the picnic will be canceled."

To complete the study of implications, NOTs are applied and the condition and conclusions are interchanged.

Original condition	$p \longrightarrow q$	
Converse	$q \longrightarrow p$	Interchange the condition and conclusion.
Inverse	$\sim p \longrightarrow \sim q$	Negate the condition and conclusion.
Contrapositive	$\sim q \longrightarrow \sim p$	Interchange and negate the condition and conclusion.
Biconditional	$(p \longrightarrow q) \wedge (q \longrightarrow p)$	Symbolized by $p \longleftrightarrow q$.

The truth values of these variations are defined as:

p	q	$\sim p$	$\sim q$	$p \to q$	Converse $q \to p$	Inverse $\sim p \to \sim q$	Contra-positive $\sim q \to \sim p$	Bi-conditional $p \longleftrightarrow q$
T	T	F	F	T	T	T	T	T
T	F	F	T	F	T	T	F	F
F	T	T	F	T	F	F	T	F
F	F	T	T	T	T	T	T	T

Notice that $p \to q$ and $\sim q \to \sim p$ are equivalent, as are $\sim p \to \sim q$ and $q \to p$. The general rule for negating an implication is to interchange and negate the conditon and conclusion to keep the statement logically equivalent.

Example

"If you live in Boston, you live in Massachusetts" is logically equivalent to "If you do not live in Massachusetts, you do not live in Boston."

The biconditional is a very strong statement. An example of a biconditional is "If the square of the longer side of a triangle is equal to the sum of the squares of the other two sides, the triangle is a right triangle" and "If a triangle is a right triangle, the square of the hypotenuse is equal to the sum of the squares of the other two sides." *If and only if* are the words used to signify a biconditional: "The square of the longer side of a triangle is equal to the sum of the squares of the other two sides if and only if the triangle is a right triangle."

EXERCISES

1. Write the following statement in if-then format.
   a. $x + 3 = 7$ if $x = 4$.
   b. You pass the test only if you know the material.
   c. For a candidate to win, it is sufficient that he or she receive a majority of votes.
   d. Spirits soar when the sun shines.
2. Write the following statements in if-then format.
   a. The crops survived only if the plants were protected.
   b. $16 = 24 - x$ if $16 + x = 24$.
   c. Politicians make promises when they are campaigning.
   d. For a student to earn a credit in a course, it is sufficient for the student to demonstrate good knowledge of the material.
3. Translate the sentences in Exercise 1 into symbolic logic, indicating the simple sentence each letter represents.
4. Translate the sentences in Exercise 2 into symbolic logic, indicating the simple sentence each letter represents.

5. Let p represent "You respect your parents" and q represent "You enjoy school." Translate each of the following:
   a. $p \to \sim q$
   b. $\sim q \to \sim p$
   c. $(q \wedge \sim q) \to p$
   d. $p \vee (q \to p)$
6. Let p represent "You like to ski" and q represent "You play the piano." Translate each of the following:
   a. $p \to \sim q$
   b. $\sim q \to \sim p$
   c. $(q \vee \sim q) \to p$
   d. $p \wedge (p \to q)$
7. Write the converse, inverse, contrapositive, and biconditional for each statement.
   a. If you exceed the speed limit, then you get a ticket.
   b. If two sides of a triangle are equal, then the triangle is isosceles.
   c. $p \to \sim q$
8. Write the converse, inverse, contrapositive, and biconditional for each statement.
   a. If you write a bad check, the bank will charge a fee.
   b. If $x = 3$, then $x - 1 = 2$.
   c. $\sim q \to \sim p$
9. Write the truth table for each of the following:
   a. $(p \to \sim q) \vee p$
   b. $(p \to q) \vee (\sim p \to q)$
   c. $(p \wedge q) \leftrightarrow (p \vee q)$
10. Write the truth table for each of the following:
   a. $p \wedge (p \to \sim q)$
   b. $(\sim p \to q) \vee (p \to \sim q)$
   c. $(p \vee q) \to (p \wedge q)$
11. Rewrite each of the compound statements in Exercise 5 in a different, but equivalent, way.
12. Rewrite each of the compound statements in Exercise 6 in a different, but equivalent, way.

ANSWERS

2. a. If the crops survived, then the plants were protected.
   b. If $16 + x = 24$, then $16 = 24 - x$.
   c. If they are campaigning, then politicians make promises.
   d. If a student demonstrates a good knowledge of material, then the student earns a credit in the course.
4. a. p: The crops survive; q: The plants were protected; $p \to q$
   b. p: $16 + x = 24$; q: $16 = 24 - x$; $p \to q$

c. p: Politicians compaign; q: Politicians make promises; $p \to q$
d. p: A student demonstrates a good knowledge of material; q: A student earns a credit in the course; $p \to q$

6. a. If you like to ski, you do not play the piano.
b. If you do not play the piano, you do not like to ski.
c. If you play the piano or not, you like to ski.
d. You like to ski, and if you like to ski you play the piano.

8. a. Converse: If the bank charges you an overdraw fee, you wrote a bad check.
Inverse: If you do not write a bad check, the bank will not charge an overdraw fee.
Contrapositive: If the bank does not charge you an overdraw fee, you do not write a bad check.
Biconditional: You write a bad check if and only if the bank charges you an overdraw fee.
b. Converse: If $x - 1 = 2$, then $x = 3$.
Inverse: If $x \neq 3$, then $x - 1 \neq 2$.
Contrapositive: If $x - 1 \neq 2$, then $x \neq 3$.
Biconditional: $x = 3$ if and only if $x - 1 = 2$.
c. Converse: $\sim p \to \sim q$
Inverse: $q \to p$
Contrapositive: $p \to q$
Biconditional: $\sim p \longleftrightarrow \sim q$ or $\sim q \longleftrightarrow \sim p$

10. a.

p	q	$\sim q$	$p \to \sim q$	$p \wedge (p \to \sim q)$
T	T	F	F	F
T	F	T	T	T
F	T	F	T	F
F	F	T	T	F

b.

p	q	$\sim p$	$\sim q$	$\sim p \to q$	$p \to \sim q$	$(\sim p \to q) \vee (p \to \sim q)$
T	T	F	F	T	F	T
T	F	F	T	T	T	T
F	T	T	F	T	T	T
F	F	T	T	F	T	T

c.

p	q	$p \vee q$	$p \wedge q$	$(p \vee q) \to (p \wedge q)$
T	T	T	T	T
T	F	T	F	F
F	T	T	F	F
F	F	F	F	T

12. a. If you play the piano, then you do not like to ski.
b. If you like to ski, you like to play the piano.
c. You like to ski.
d. You like to ski and if you do not like to play the piano, then you do not like to ski.

3.4 SIMPLIFICATION OF LOGIC STATEMENTS

Often a verbal statement is so lengthy and complicated that it is difficult to ascertain the true meaning. Simplifying statements can often clarify their meanings.

DeMorgan's rules are tools for simplifying a compound statement preceded by *it is not true that.* When a negation symbol precedes parentheses enclosing a compound statement, DeMorgan's rules state that each part of the compound statement is negated and the connective is changed—AND becomes OR, and vice versa. DeMorgan's rules apply only to compound statements with AND and OR as connectives.

Example

$$\sim(p \vee \sim q) = \sim p \wedge q$$

$$\sim(\sim p \wedge q) = p \vee \sim q$$

DeMorgan's rules apply to symbolized AND and OR compound statements preceded by a negation and verbal compound AND and OR statements preceded by *it is not true that.*

Example

"It is not the case that the employee is over 50 and has worked more than 5 years" simplifies to "The employee is not over 50 or has not worked more than 5 years."

DeMorgan's rules can be verified by truth tables. Exercises 5 and 7 and 4 and 6 in Section 3.2 demonstrate the equivalences of the rules.

Symbolic logic is a branch of mathematics; the properties of mathematics associated with numbers and sets also apply in logic:

1. Commutative properties: $p \vee q = q \vee p; p \wedge q = q \wedge p$
2. Associative properties: $p \vee (q \vee r) = (p \vee q) \vee r;$ $p \wedge (q \wedge r) = (p \wedge q) \wedge r$
3. Distributive properties: $p \wedge (q \vee r) = (p \wedge q) \vee (p \wedge r);$ $p \vee (q \wedge r) = (p \vee q) \wedge (p \vee r)$

$$p \vee \sim p = T \qquad p \wedge \sim p = F \qquad \sim(\sim p) = p$$

When no parentheses occur in a logic statement to indicate what connective takes precedent, hierarchy of operations dictates that NOT connectives are applied first and affect only the statements directly following NOTs. AND is the next connective to be applied, and OR is the last connective used. If more than one of a given connective is present, work from left to right, picking up all of the given connective. If-then statements are the same level as OR statements, and so are considered after NOT and AND.

Some examples illustrate the use of the properties of logic, the hierarchy of operations, and DeMorgan's rules.

Example

Simplify $p \vee \sim p \wedge q$. No grouping symbols exist, so the hierarchy of connectives must be applied. The subscripts following the parentheses indicate the order the connectives must be considered.

$$(p \vee ((\sim p)_1 \wedge q)_2)_3$$

Use the distributive property to rewrite the above as:

$$(p \vee \sim p) \wedge (p \vee q)$$

$p \vee \sim p$ is always true; it is a tautology.

$$T \wedge (p \vee q) = p \vee q$$

This same result can be achieved by constructing a truth table for $p \vee \sim p \wedge q$

p	q	$\sim p$	$\sim p \wedge q$	$p \vee (\sim p \wedge q)$	$p \vee q$
T	T	F	T	T	T
T	F	F	F	T	T
F	T	T	T	T	T
F	F	T	F	F	F

Logically equivalent

Example

Simplify $\sim(p \wedge \sim q) \wedge (\sim p \vee q)$

$(\sim p \vee q) \wedge (\sim p \vee q)$ Apply DeMorgan's rules;

$(\sim p \vee q)$

Example

Simplify P AND (Q EOR NOT R). Recall p EOR $q = (p \wedge \sim q) \vee (\sim p \wedge q)$.

$P \wedge ((Q \wedge R) \vee (\sim Q \wedge \sim R))$

$(P \wedge Q \wedge R) \vee (P \wedge \sim Q \wedge \sim R)$ Distributive property

Example

Simplify $\sim p \rightarrow r \vee (q \vee r)$

$p \vee r \vee q \vee r$ Replace $\longrightarrow$ by OR.

$p \vee q \vee r$

Example

Simplify $p \rightarrow (\sim p \rightarrow q)$.

$$p \rightarrow (p \vee q)$$

$$\sim p \vee p \vee q$$

$$(\sim p \vee p) \vee q$$

This is a tautology.

EXERCISE

Simplify using rules of simplification.

1. NOT (p OR NOT q)
2. p OR NOT (NOT p AND q)
3. NOT $p \rightarrow$ (p AND NOT q)
4. p AND (NOT p EOR q)
5. NOT (NOT p AND (p OR NOT q))
6. $p \rightarrow$ ($q \rightarrow$ NOT p)
7. NOT (NOT p AND q) AND p OR NOT q
8. If (NOT P AND Q), then Q
9. p OR q AND NOT p
10. NOT p AND q OR p

Symbolize the following and simplify if possible. Use the indicated letters.

11. Two sides of a triangle are equal if and only if the angles opposite these sides are equal (S, A).
12. If the major factors of crime are high unemployment, boredom, and crowded living conditions, more jobs and better living conditions are needed (H, B, C, J, N).
13. If you sold your residence in the past year, you must pay income tax on the profits or reinvest the profit within 12 months (S, P, R).
14. The present trend toward nuclear disaster cannot be averted if (a) negotiations are not held, (b) the arms race accelerates, (c) the people do not demand steps to avert the disaster (D, N, R, P).
15. The Braves win only if the Dodgers lose (B, D).

Answers

2. p OR NOT q
4. p AND q
6. NOT q or NOT p
8. A tautology

10. p OR q
12. $(H \wedge B \wedge C) \rightarrow (J \wedge N)$
14. $(\sim N \wedge R \wedge \sim P) \rightarrow \sim D$

3.5 FLOWCHARTS

In order to use a computer to solve a problem, information must be presented in an orderly and sequential manner. A flowchart is a picture of this orderly and sequential handling of information and decision-making. Each of the following shapes suggests a particular function in a flowchart.

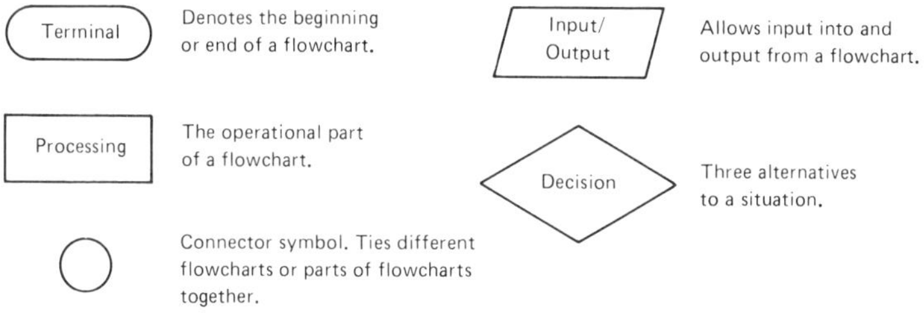

Generally, the action described by a flowchart moves down and right through the diagram. If other directions are involved, arrows must be present to direct the flow.

In a decision box, the diagram flows into the decision box on the upper vertex. If the decision is yes or true, the flow is from the lower vertex. If the decision is no or false, the flow is usually from the rightmost vertex, unless this creates a crossing of flow lines; then the leftmost vertex is used to continue the diagram. If the vertices are not labeled, the lower one is the positive, or yes, terminal. The shape of the decision box also allows handling three alternatives to a decision. Suppose the decision is to compare two numbers, a and b. If $a < b$, the flow can be from the left vertex; if $a = b$, from the lower vertex; and if $a > b$, from the right vertex.

Example

You require a list of all employees who have worked over 10 years and are 60 years of age or older.

Every decision box must have two lines from it to indicate the path in each situation.

The no vertex is on the right for the first and second decision box and on the left for the last decision box. This is to avoid the flow lines from intersecting.

The decision boxes concerning years of service and age could have been interchanged. Usually the decision that eliminates the largest group is made first.

The arrangement of the decision boxes in a flowchart reflects the connective in the original problem.

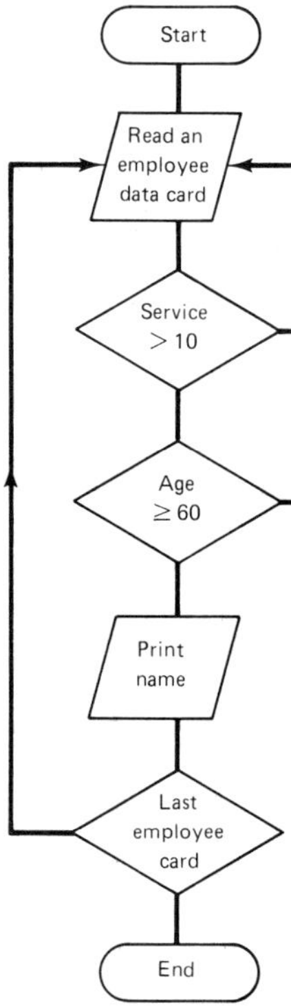

Although the entire flowchart is necessary and useful to orderly and sequential thinking, the arrangement of the decision boxes is the tie to logic and the prime concern of this section.

Decision boxes can be arranged in two ways.

TRUTH TABLE FOR p AND q

p	q	$p \wedge q$
1. T	T	T
2. T	F	F
3. F	T	F
4. F	F	F

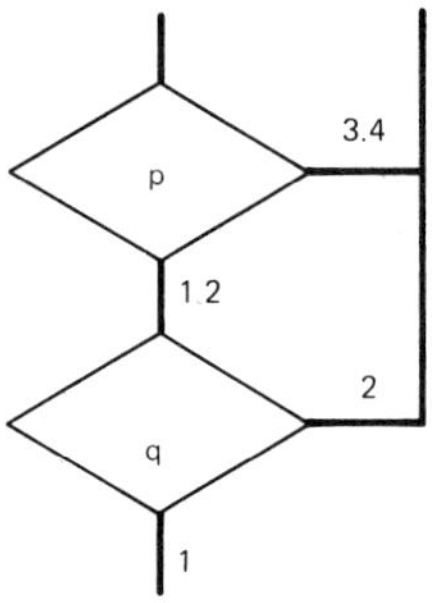

The numbers on the flowchart correspond to the lines in the truth table. Only line 1 of the truth table is T. Therefore, only that condition allows flow through the diagram.

TRUTH TABLE FOR *p* OR *q*

	p	q	$p \vee q$
1.	T	T	T
2.	T	F	T
3.	F	T	T
4.	F	F	F

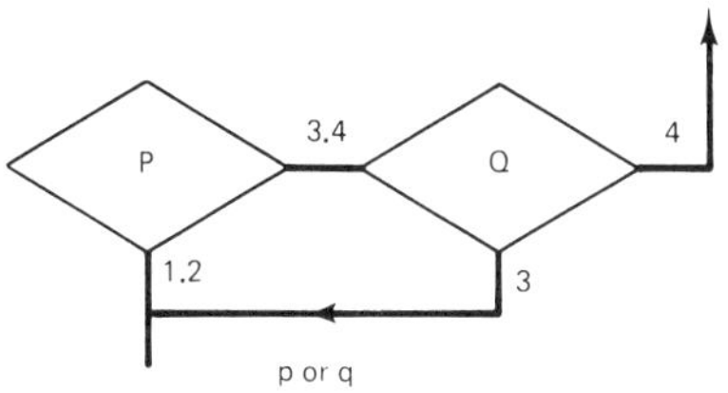

p or q

The numbers on the flowchart correspond to the lines in the truth table. Lines 1, 2, and 3 are T and the flowchart indicates that these conditions allow flow through the diagram.

If the connective NOT is a part of a decision, the simple statement is the decision, and the outputs on the vertices are interchanged to indicate NOT. If the vertices are not labeled, assume the lower one to be T.

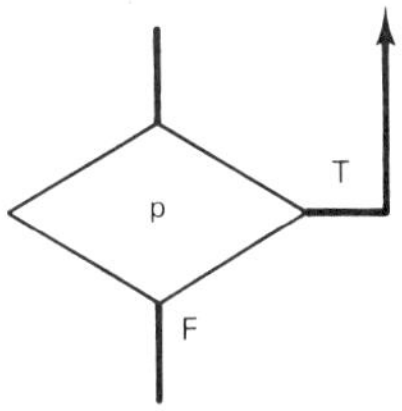

Not p

Example

Flowchart (p or not q). Because the connective is OR, the boxes are horizontal. Because we want NOT q, the outputs on the q decision box are interchanged.

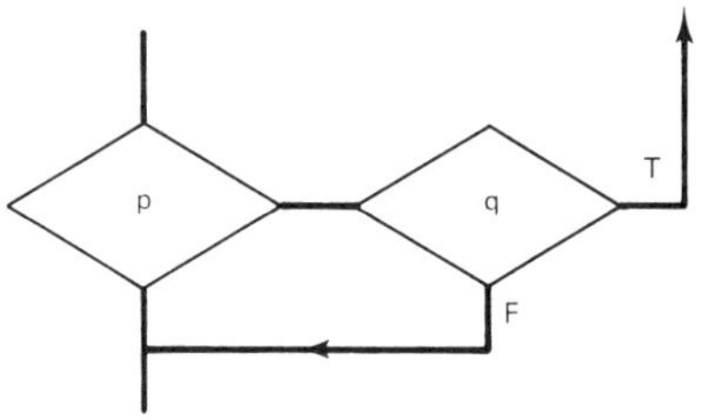

p or not q

Most problems involve several decisions. Before flowcharting, the problem should be simplified and clarified. This is done by constructing a truth table or utilizing simplification techniques.

Another type of statement that can be flowcharted is the conditional. If a decision is yes, the flow continues in a vertical path. If a decision is no, the flow is horizontal to a different set of instructions. Eventually the two paths meet and proceed in the flowchart.

Example

If your last name begins with letters *A* through *L*, go to station *A*, station *B*, and station *C*. If your last name does not begin with letters *A* through *L*, go to station *D*, station *E*, and station *C*.

A flowchart for this situation is:

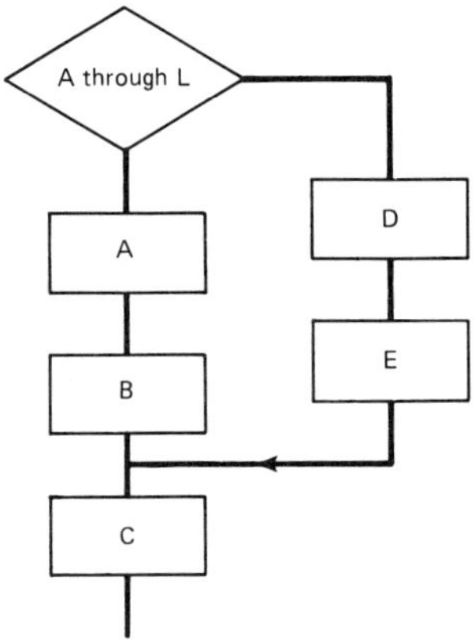

Example

If *A* and *B*

C

D

else *F*

G.

E

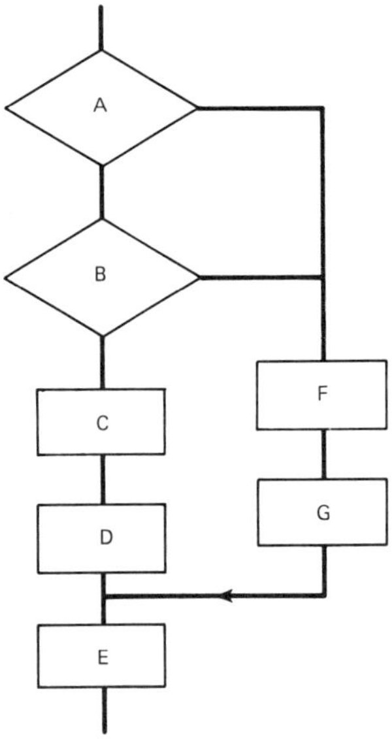

These instructions indicate to check decision boxes A and B. If both are yes (because of AND), go to processing boxes C, D, and G. If NOT (A and B), go to processing boxes E, F, and G. The word *else* is the key for the NOT (A and B). The *period* indicates where the two routes flow back together again.

EXERCISES

Indicate the decision-box arrangement and flowchart for the following. Simplify the expressions where possible before flowcharting.

1. NOT p AND q
2. p AND NOT q
3. NOT p OR q
4. NOT p OR NOT q
5. p EOR NOT q
6. NOT p EOR q
7. $p \rightarrow (q \rightarrow \text{NOT } p)$
8. $(p \text{ OR } q) \rightarrow \text{NOT } r$
9. NOT (NOT p AND q) AND p
10. p OR NOT (q AND NOT r)
11. NOT A AND B OR D AND E
12. A OR NOT B AND C

Indicate the decision-box arrangement and instruction boxes for the following:

13. A and not B
    C
    D
    else
    E.
    F
14. A or not B
    C
    else
    D
    E.
    F

Write a flowchart for the following truth table outputs. Indicate where each line of the truth table is on the flowchart.

15.

	p	q	Output
1.	T	T	F
2.	T	F	T
3.	F	T	T
4.	F	F	F

16.

	p	q	Output
1.	T	T	T
2.	T	F	F
3.	F	T	F
4.	F	F	T

17. Write a flowchart for the following situation. B. C. Smith Dept. store is sending out monthly statements to its customers. If the bill is over $25, indicate the total amount, add carrying charges, and state the amount due is one-third the total amount. Otherwise, state the amount due is the total amount. In both cases, print the bill.
18. Government regulations demand a list of all employees who worked around asbestos and have been hospitalized during the past 12 months. Write the flowchart that would determine the list.
19. A college desires a printout of students on the dean's list. To be eligible for the dean's list, a student must have carried at least 12 credit hours, achieved a GPR (grade point ratio) of at least 3.5 and had no grade less than a *C* in that semester. Write the flowchart to determine the dean's list.
20. The annual bonus plan in one company is determined by annual salary. If the annual salary is $10,000 or more, the bonus is $500. Otherwise, the bonus is 5% of the annual salary. Make a record of the bonus due each employee. Write the flowchart to determine the bonuses.

Answers

2.

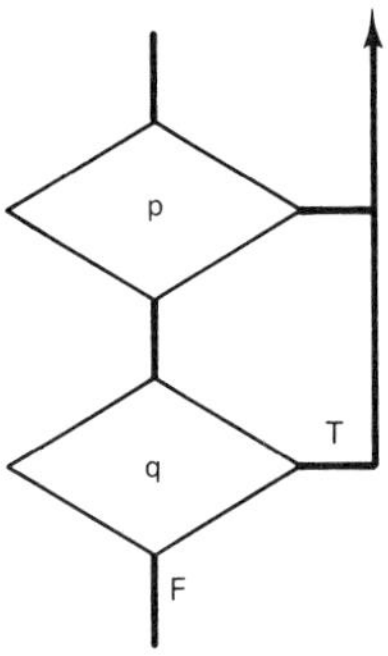

4.

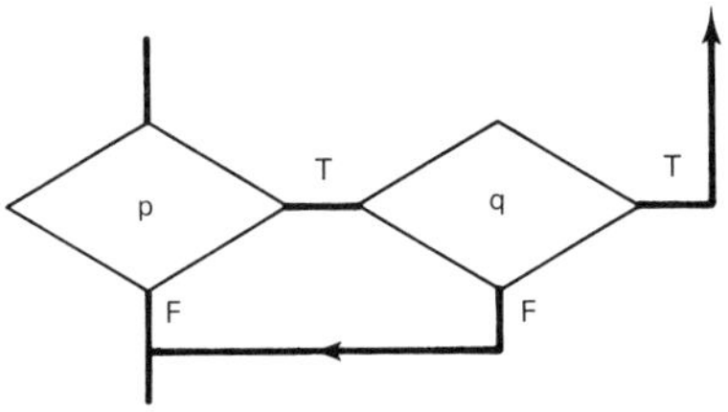

6.

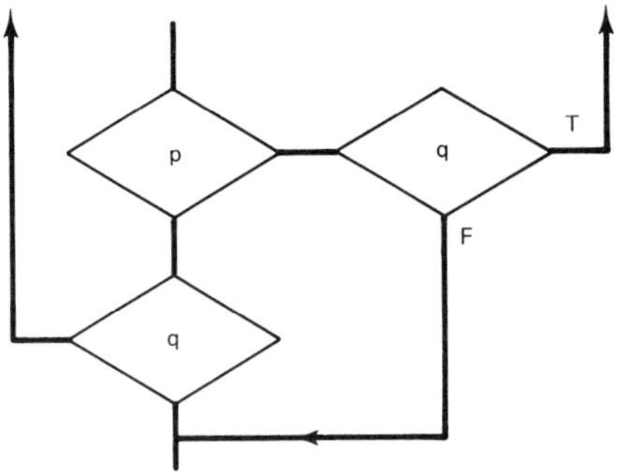

8.

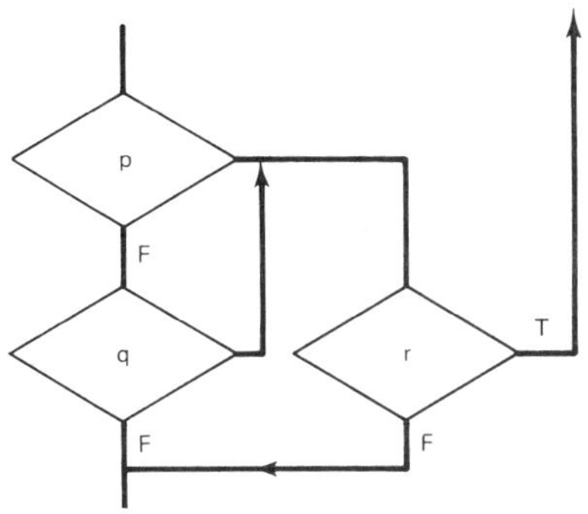

10.

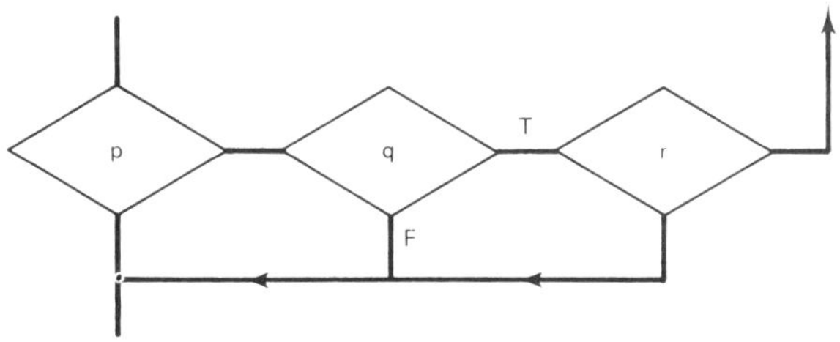

12.

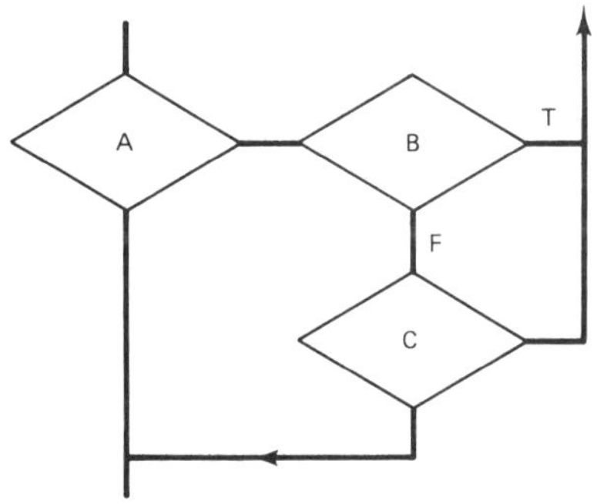

14.

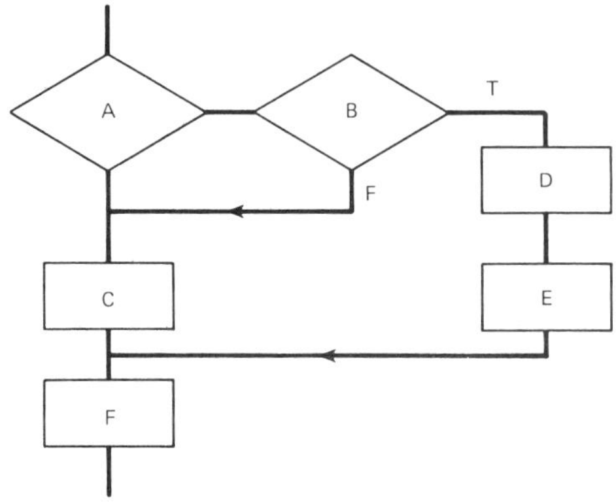

16.

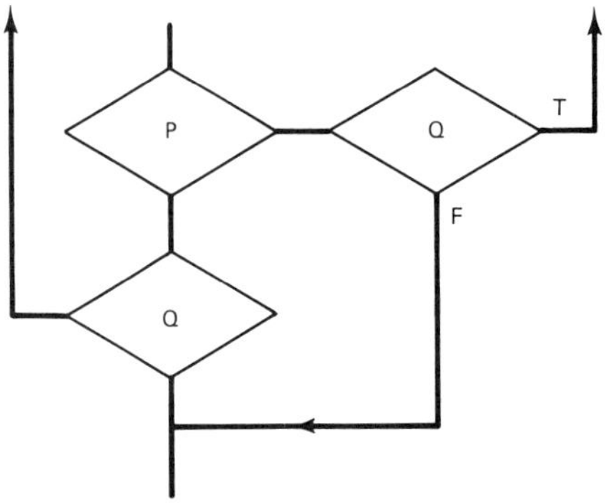

18.

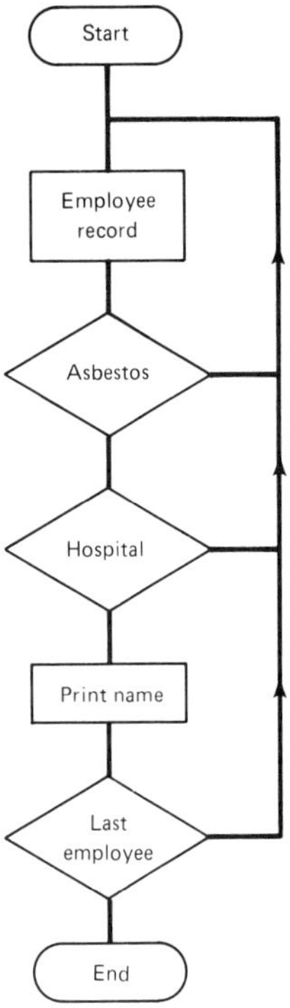

20.

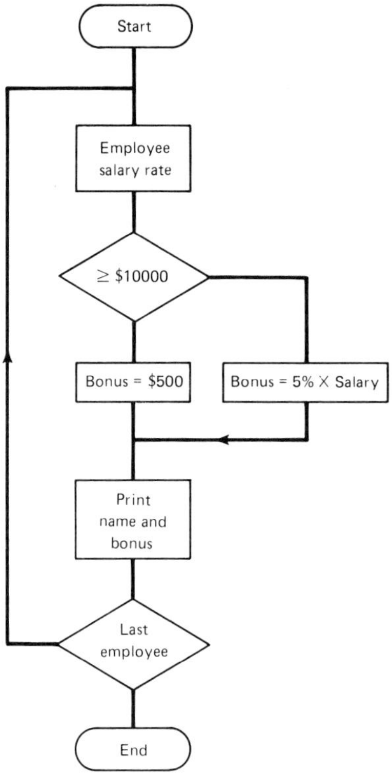

CHAPTER 3 REVIEW EXERCISES

Symbolize the following statements using the following representations:

p = George gets a promotion q = Jim gets a raise r = Pete gets a transfer

1. George gets a promotion or Pete gets transferred, but not both.
2. If Jim gets a raise, George does not get a promotion.
3. Pete gets a transfer only if George gets a promotion.
4. Pete gets a transfer, Jim gets a raise, and George does not get a promotion.
5. It is not true that Jim did not get a raise or George got a promotion.
6. George gets a promotion or Jim gets a raise, and Pete gets a transfer.
7. George gets a promotion if and only if Pete gets a transfer.
8. George gets a promotion when Jim gets a raise.

Construct a truth table for each of the following:

9. $p \wedge \sim q$
10. $\sim p \vee q$
11. $\sim p \rightarrow (q \wedge p)$
12. $p \vee \sim(p \vee q)$
13. NOT p EOR q
14. (NOT p AND q) $\rightarrow q$
15. p OR q AND NOT p
16. NOT (NOT p OR q) OR (NOT p AND q)

Verbalize the following if p represents "An employee earns over \$20,000 a year," q represents "An employee is at least 50 years old," and r represents "An employee has worked at least 15 years with the company."

17. $p \wedge q \wedge \sim r$
18. $\sim p \vee \sim q$
19. q EOR p
20. $p \vee \sim q \vee \sim r$
21. $\sim q \rightarrow (\sim r \wedge \sim p)$
22. $\sim(p \wedge q)$
23. Given the statement: "If the war on inflation is not won, the recession will continue."
    a. Write the inverse.
    b. Write the converse.
    c. Write the contrapositive.
    d. Indicate which statements are logically equivalent.
24. Given: "If students have a strong background in mathematics, opportunities in high-technology areas will be abundant."
    a. Write the inverse.
    b. Write the converse.
    c. Write the contrapositive.
    d. Indicate which statements are logically equivalent.

Draw the arrangement of the decision boxes for each of the following:

25. NOT p EOR NOT q
26. p AND NOT q OR r

27. $p \rightarrow (q$ AND NOT $r)$
28. NOT $(p$ AND $r)$ OR q

Draw the flowchart for each of the following conditionals:

29. If A or not B and not C

    D

    else

    E.

    F

30. If not A and B or C

    D

    E

    else

    F.

    G

Chapter 3 Review Answers

2. $q \rightarrow \sim p$
4. $r \wedge q \wedge \sim p$
6. $(p \vee q) \wedge r$
8. $q \rightarrow p$
10. *TFTT*
12. *TTFT*
14. *TTTT*
16. *FTTF*
18. An employee does not earn \$20,000 or is not at least 50 years old.
20. An employee earns over \$20,000, is not 50 years old, or has worked less than 15 years with the company.
22. It is not true an employee earns over \$20,000 and is at least 50 years old.
24. a. If students do not have a strong background in mathematics, opportunities in high-technology areas will not be abundant.
    b. If opportunities in the high-technology areas are abundant, students do have a strong background in mathematics.
    c. If opportunities in high-technology areas are not abundant, students do not have a strong background in mathematics.
    d. (a) and (b) are equivalent; (c) and the original are equivalent.

26.

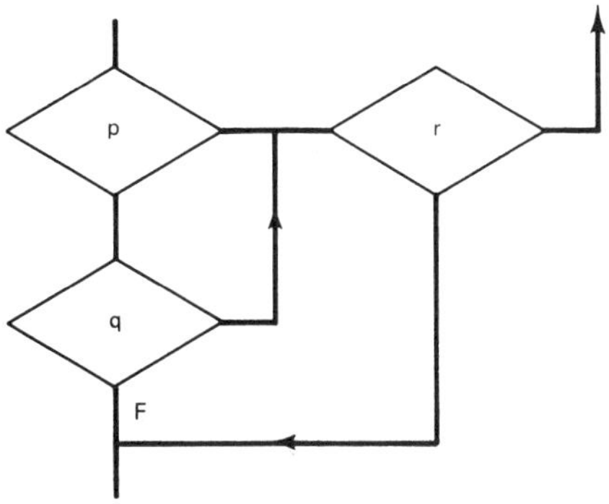

28.

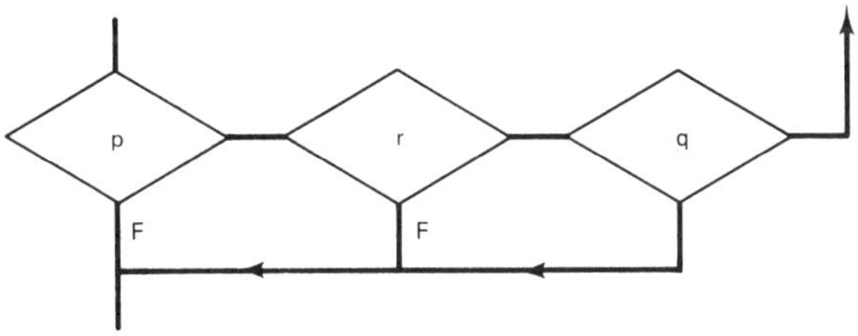

Boolean Algebra

CHAPTER 4

4.0 INTRODUCTION

Boolean algebra refers to a branch of mathematics developed by George Boole over 100 years ago. The algebra formalized a relationship between logic and mathematics. Boole took the concepts of logic and applied some of the symbolism of algebra to produce a new branch of mathematics. This development was an academic exercise; it was dormant until the advent of electricity and electronics. The set of properties formalized by Boole provided the necessary mathematics for circuit design.

A study of Boolean algebra includes a brief introduction to switching circuits in electricity. Concepts involving switching circuits, logic, and algebra are combined into a set of properties and notation called *Boolean algebra.* Computers are masses of integrated circuits; their functioning and design are dependent on the concepts introduced in this chapter.

The material in this chapter is an introduction. Additional techniques are available to assist in the design and simplification of circuits.

This chapter is designed to:

1. Provide the notation and properties of Boolean algebra.
2. Provide the techniques involved in using properties to simplify Boolean expressions.
3. Provide the techniques needed to translate from circuit notation to Boolean algebra and from Boolean algebra notation to circuit notation.

4.1 BASIC CIRCUITS

The basic component of a circuit is a switch. A switch determines the output of the circuit. A switch can be closed (on) or open (off). The schematics to represent these two states are as follows:

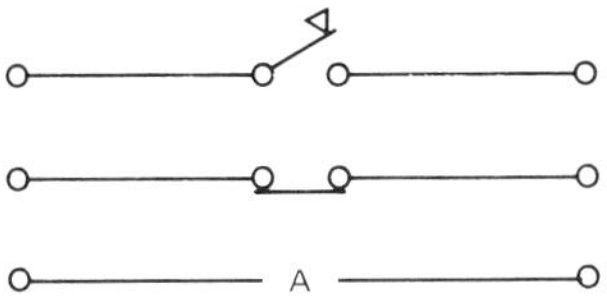

Switch is in the OFF position; no current flows.

Switch is in the ON position; current flows.

General representation of a switch A, whose state may vary.

Two or more switches can be arranged in a variety of ways in complex circuits. However, all complex circuits are combinations of two basic circuits, parallel and series.

Series connections. *Series connections* refer to switches that are connected so current flows only when all the switches are closed. The schematic to illustrate two switches connected in series is as follows:

Since each switch can be either on or off, there are four possible combinations involved. Here, 0 represents an open switch, where no current can flow, and 1 represents a closed switch, where current can flow. A diagram representing the four possible combinations is:

Switch A	Switch B	Output	Series Notation
1	1	1	$1 \cdot 1 = 1$
1	0	0	$1 \cdot 0 = 0$
0	1	0	$0 \cdot 1 = 0$
0	0	0	$0 \cdot 0 = 0$

The symbol denoting switches connected in series is $\cdot$ (like the multiplication sign). Thus $A \cdot B$ indicates that switches A and B are connected in series. As in mathematics, the dot is usually omitted and the letters are written next to each other. Hence, the notation to show that switches A and B are connected in series is:

$$A \cdot B = AB$$

The series notation for the four possible combinations of outputs is listed separately as

$$1 \cdot 1 = 1$$

$$1 \cdot 0 = 0$$

$$0 \cdot 1 = 0$$

$$0 \cdot 0 = 0$$

The table is consistent with the multiplication facts for binary and decimal notation.

Parallel connections. *Parallel connections* refer to switches that are connected so current flows when any or all switches are on. The schematic to illustrate two switches connected in parallel is as follows:

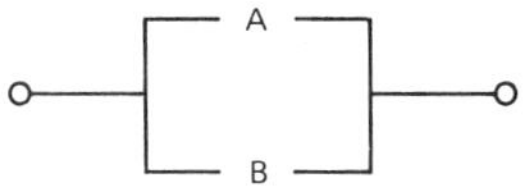

The verbal description of *parallel* seems appropriate from the schematic. The switches are on parallel lines. Since each switch can be on or off, there are four possible combinations involved. Again, 0 represents an open switch, where no current can flow, and 1 represents a closed switch, where current flows. A diagram representing the four possible combinations is:

Switch A	Switch B	Output	Parallel Notation
1	1	1	$1 + 1 = 1$
1	0	1	$1 + 0 = 1$
0	1	1	$0 + 1 = 1$
0	0	0	$0 + 0 = 0$

The symbol denoting switches connected in parallel is + (like the addition sign). Hence, $A + B$ is the notation for denoting that switches A and B are connected in parallel.

The parallel notation for the four possible combinations for output is:

$$1 + 1 = 1$$

$$1 + 0 = 1$$

$$0 + 1 = 1$$

$$0 + 0 = 0$$

This table is consistent with both binary and decimal addition, except for $1 + 1 = 1$. It is necessary to lend the concept of current to this notation to interpret this result. Current flowing through each of two parallel circuits produces an output of current.

Complement switches. *Complement switches* are two switches within a circuit such that when one is on, the other is off. These switches are not independent of each other; therefore, they do not carry different letter notation. Complementary switches guarantee the switches are in opposite states. A bar over the letter symbolizing a switch indicates the complement. Any switch has two possible states, on and off. The complement of that switch has the states of off and on. A diagram to show this relationship is:

A	$\bar{A}$
1	0
0	1

The complement in switches is similar to the NOT in logic; it reverses the value of the output.

The schematics for complements connected in series and parallel are as follows:

Current never flows.

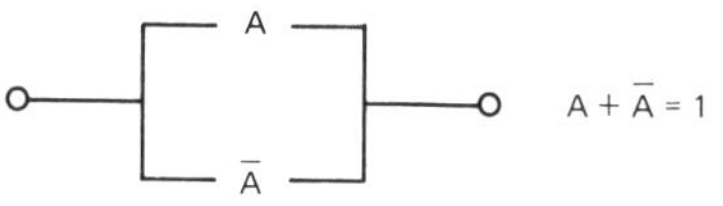

Current always flows.

The table of values for complements is:

A	$\bar{A}$	Series	Parallel
1	0	0	1
0	1	0	1

Combination circuits. *Combination circuits* are a combination of series and parallel circuits; they are called *complex circuits.*

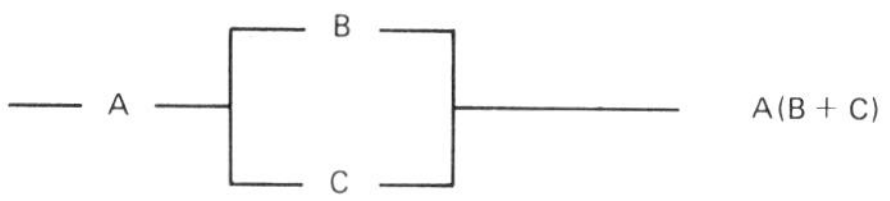

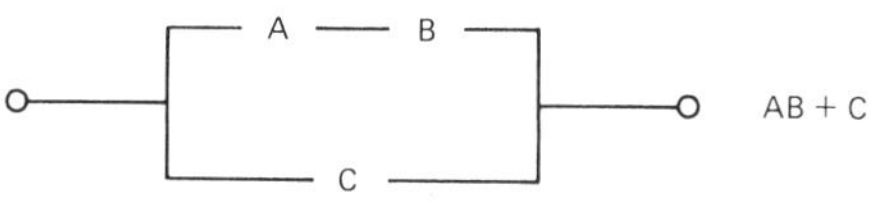

Series, parallel, and complement are analogous to AND, OR, and NOT in logic. If the notations for series, parallel, and complement used T for 1 and F for 0 and $'$ for $^{-}$, the basic truth tables of logic would emerge. The laws of symbolic logic apply to switching-circuit notation.

EXERCISES

1. Draw a circuit to represent $\bar{A} + B$. For what values of A and B is $\bar{A} + B = 1$?
2. Draw a circuit to represent $A + \bar{B}$. For what values of A and B is $A + \bar{B} = 1$?
3. Draw a circuit to represent $\bar{A} \cdot B$. For what values of A and B is $\bar{A} \cdot B' = 1$?
4. Draw a circuit to represent $A \cdot \bar{B}$. For what values of A and B is $A \cdot \bar{B} = 1$?

Find the current output of
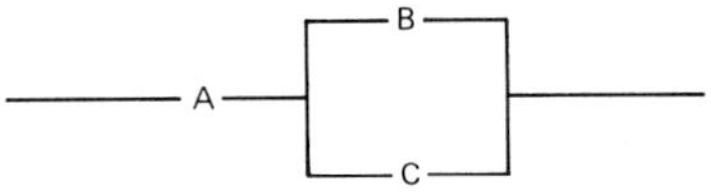

5. $A = 1, B = 1, C = 0$
6. $A = 1, B = 0, C = 1$
7. $A = 0, B = 0, C = 1$
8. $A = 0, B = 1, C = 1$

Find the current output of
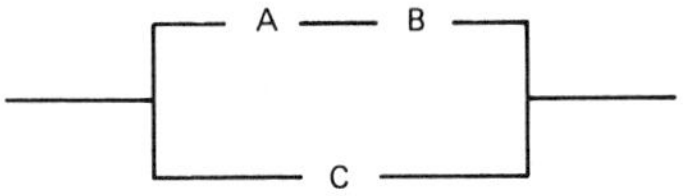

if

9. $A = 0, B = 1, C = 1$
10. $A = 1, B = 0, C = 1$
11. $A = 1, B = 0, C = 0$
12. $A = 0, B = 0, C = 1$

Find the current output of
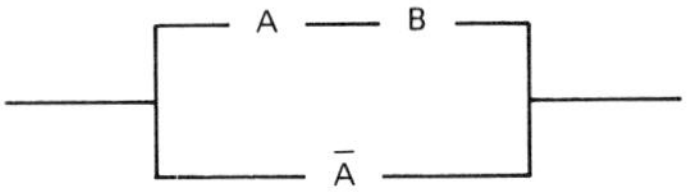

if

15. $A = 0, B = 1$
16. $A = 1, B = 0$
13. $A = 1, B = 1$
14. $A = 0, B = 0$

Answers

2.
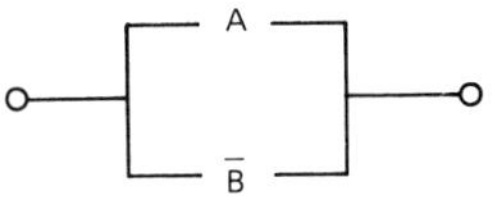

$A = 1, B = 0$; $A = 1, B = 1$; $A = 0, B = 0$

4.

$A = 1, B = 0$

6. 1 8. 0 10. 1 12. 1 14. 1 16.0

4.2 ADDITIONAL CIRCUITS

Before the properties of Boolean algebra are studied and techniques for simplifying Boolean expressions are suggested, it is necessary to be able to write

Boolean expressions to represent schematics and to draw schematics to represent Boolean expressions.

Example

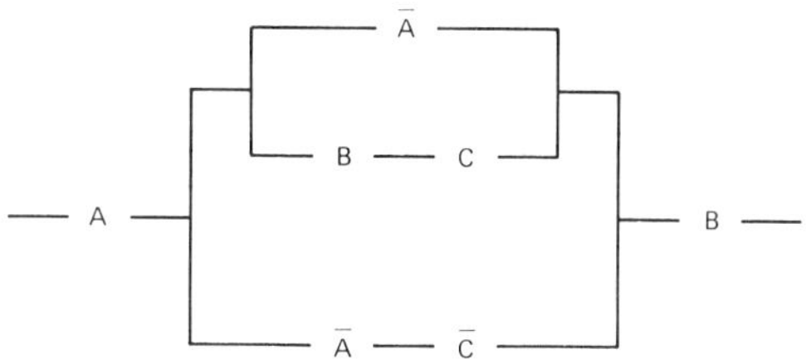

Step 1. Determine the switches in parallel: $(\bar{A} + BC)$ and $\bar{A}\bar{C}$. These can be written with no parentheses: $\bar{A} + BC + \bar{A}\bar{C}$.

Step 2. Use parentheses around the switches in parallel to denote A is in series with all these and all these are in series with B: $A(\bar{A} + BC + \bar{A}\bar{C})\,B$, which can be written $AB(\bar{A} + BC + \bar{A}\bar{C})$.

Step 3. This expression can be simplified, but that is not the task of this section.

Example

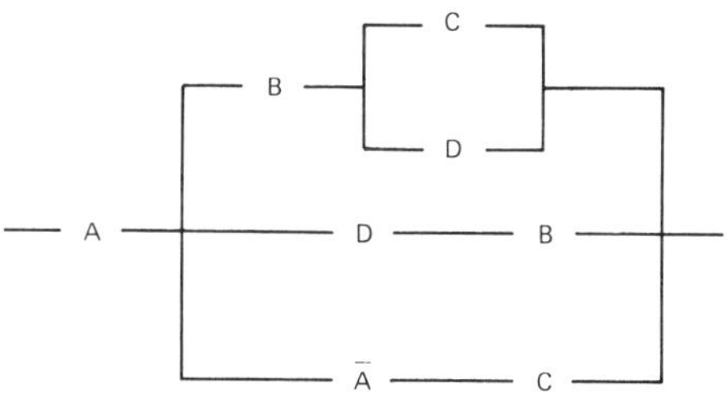

Step 1. Write each line of the parallel circuit, joining them with + line 1 = $B(C + D)$, line 2 = DB, and line 3 = $\bar{A}C$: Parallel = $B(C + D) + DB + \bar{A}C$

Step 2. Use parentheses around all of Step 1 to show it is in series with A: $A(B(C + D) + DB + \bar{A}C)$.

Step 3. Again, the task of simplification will be addressed later.

Example

Draw a schematic for $A((C + D) + B(\bar{A} + D))$.

Step 1. First, notice there are two parallel switches, $(C + D)$ and $(\bar{A} + D)$, that are connected in parallel. Draw these.

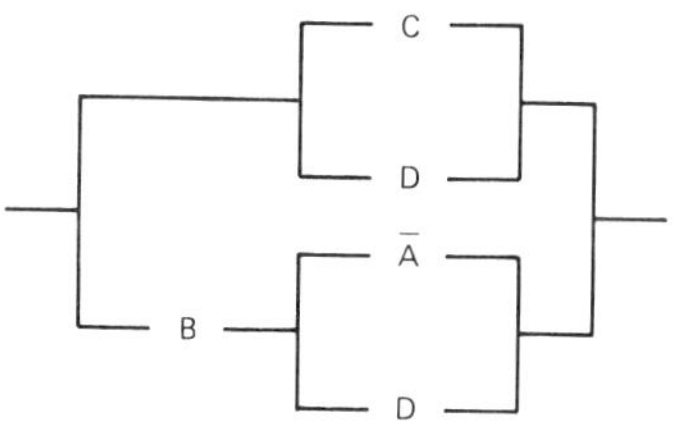

Step 2. Next, switch B is in series with $\bar{A} + D$.

Step 3. Lastly, switch A is in series with the entire diagram.

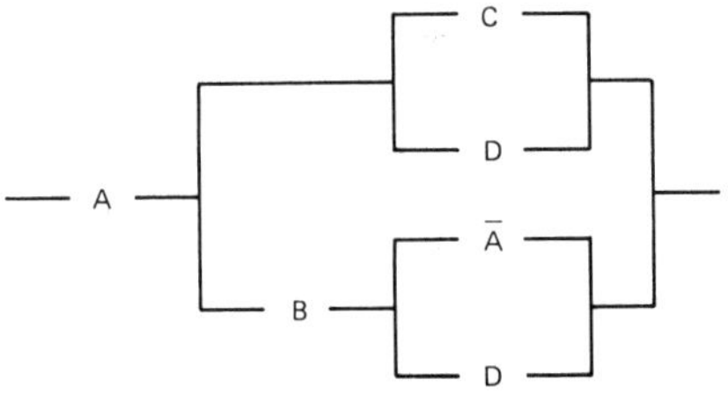

The most-troublesome points in drawing schematics or writing Boolean expressions involve the use of parentheses and the currect connectives.

EXERCISES

Draw the circuit schematic for each of the following:

1. $(A + B)(C + D)$
2. $AB + CD$
3. $A + \bar{A}B$
4. $A(C(A + \bar{A}B) + C)$
5. $(A(B + C))\,\bar{C}D$
6. $(A + \bar{C})(B + D)$
7. $A\bar{C} + BD$
8. $B + \bar{A}B$
9. $A(D(C + AD) + B)$
10. $AC(A + B + C + D)$

Write the Boolean expression for each of the following

11.

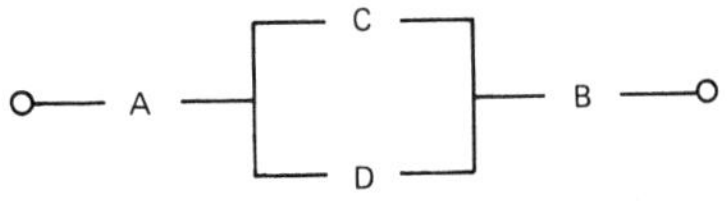

12.

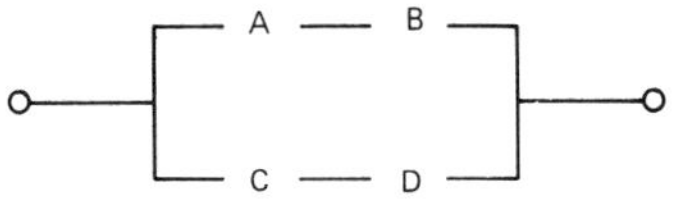

13.

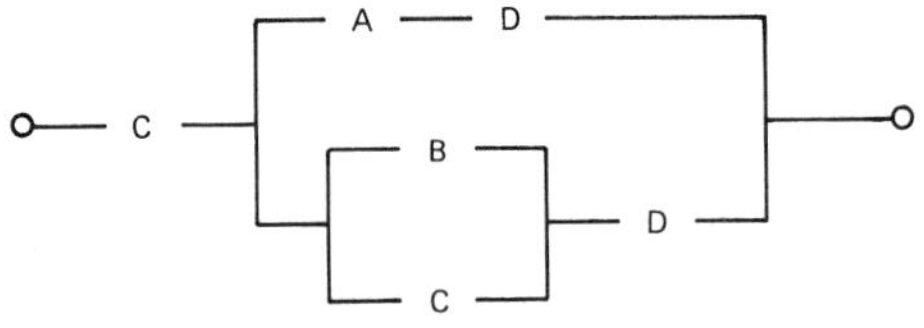

14.

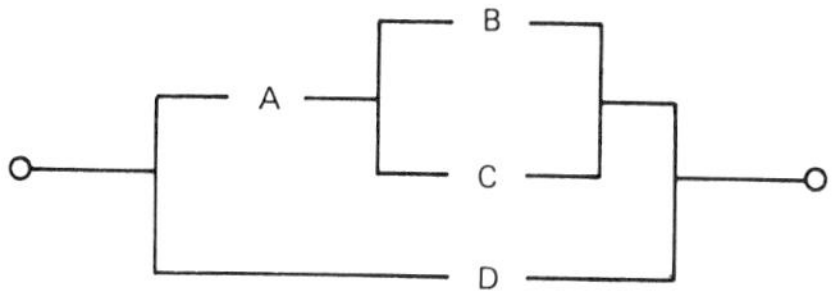

15.

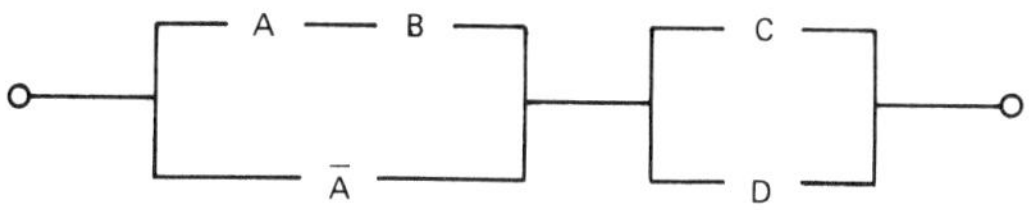

16.

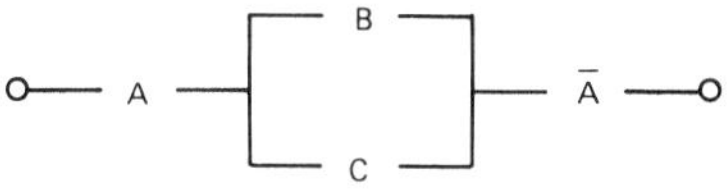

17.

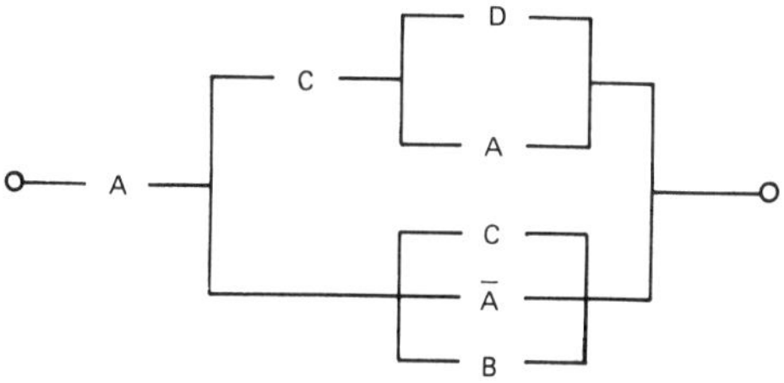

18.

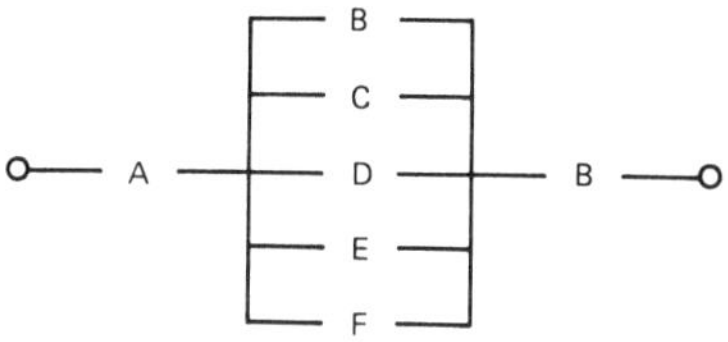

19.

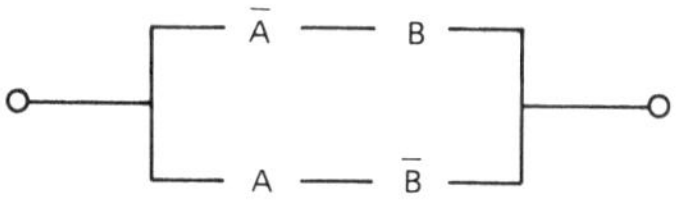

20.

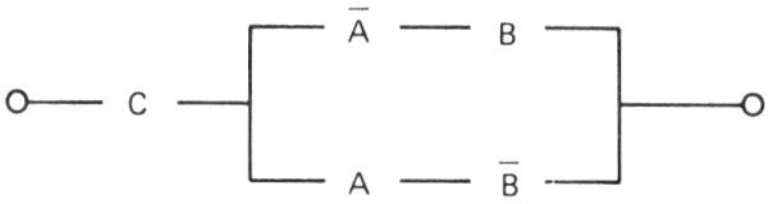

Answers

2.

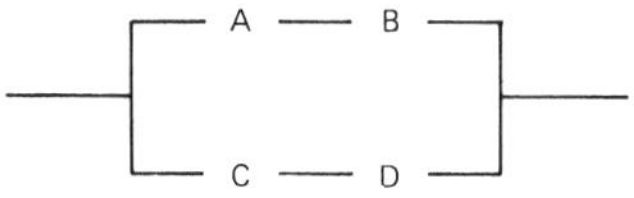

4.

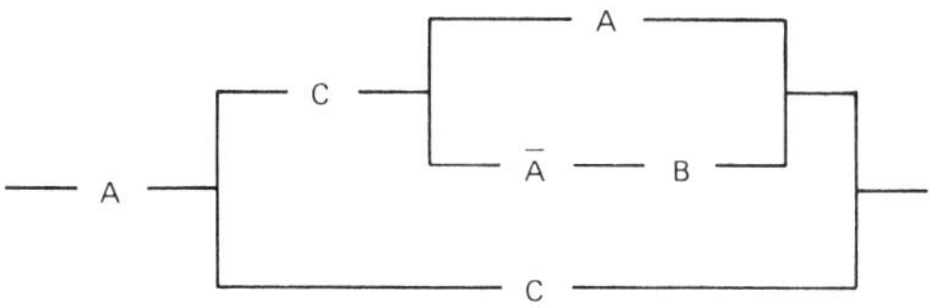

6.

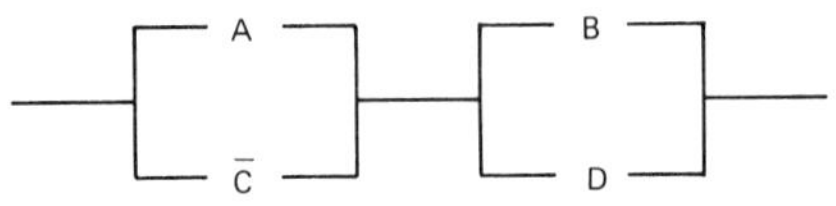

8.

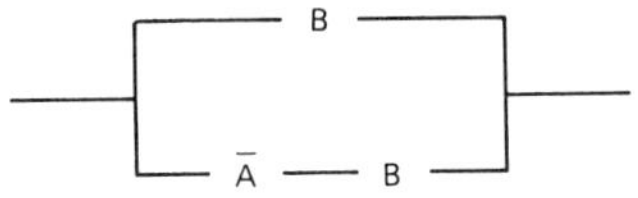

10.

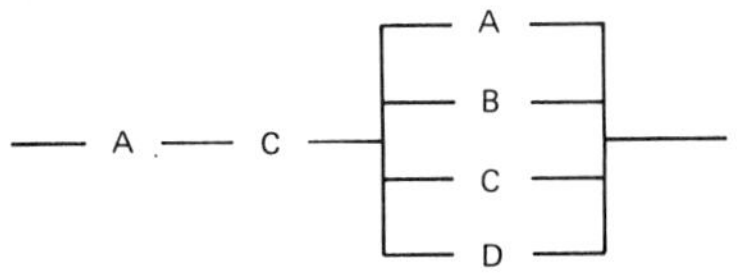

12. $AB + CD$
14. $A(B + C) + D$
16. $A\bar{A}(B + C)$
18. $AB(B + C + D + E + F)$
20. $C(\bar{A}B + A\bar{B})$

4.3 BASIC BOOLEAN PROPERTIES

In designing circuits for a particular task, it is desirable to express the design in simplest terms with the fewest number of switches. Before the advent of microchips, simplification was an economic necessity. With printed circuits, economics does not play as large a role. A simpler design has fewer troublespots.

The first ten properties listed are the properties expected in an algebra.

Commutative:	(Parallel)	1. $A + B = B + A$
	(Series)	2. $AB = BA$
Associative:	(Parallel)	3. $A + (B + C) = (A + B) + C$
	(Series)	4. $A(BC) = (AB)\,C$
Inverse:	(Parallel)	5. $A + \bar{A} = 1$
	(Series)	6. $A \cdot \bar{A} = 0$
Identity:	(Parallel)	7. $A + 0 = A$
	(Series)	8. $A \cdot 1 = A$
		9. $A + 1 = 1$
Distributive		10. $A(B + C) = AB + AC$

1. 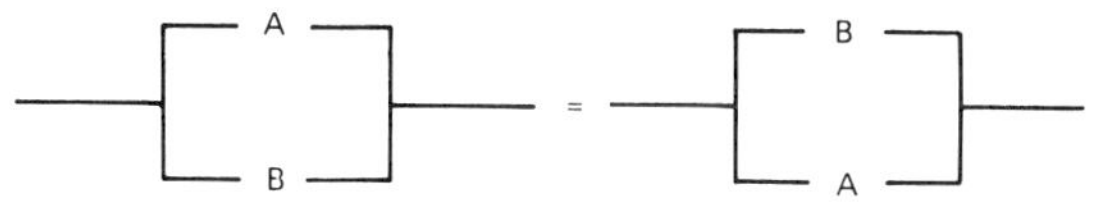

2. 

3. 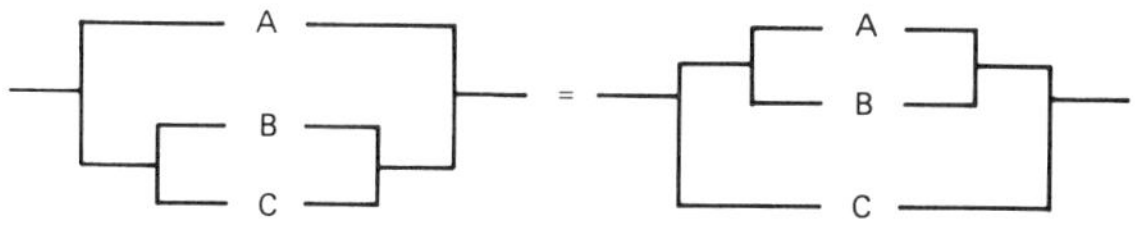

4. 

5. 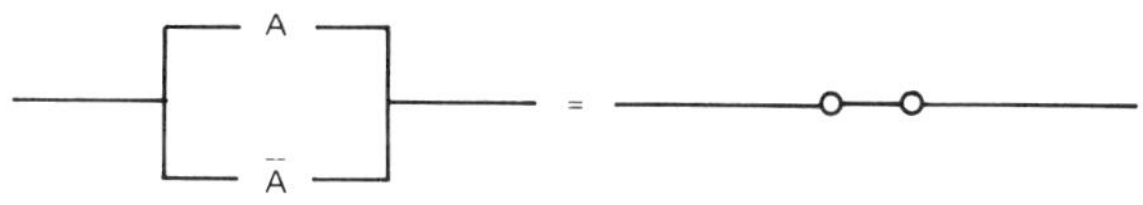

6. 

7. 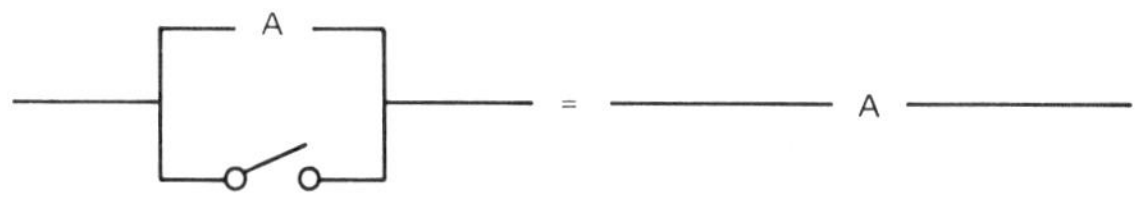

8. 

9. 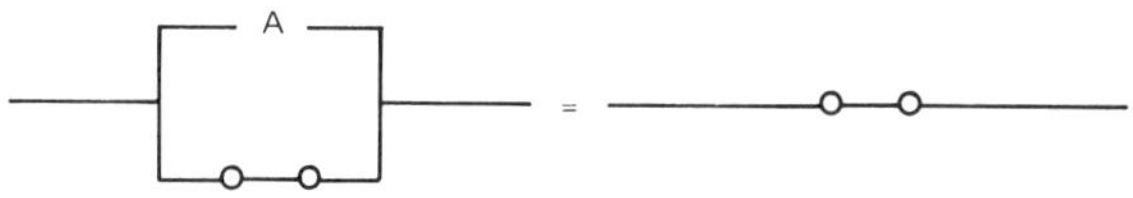

10. 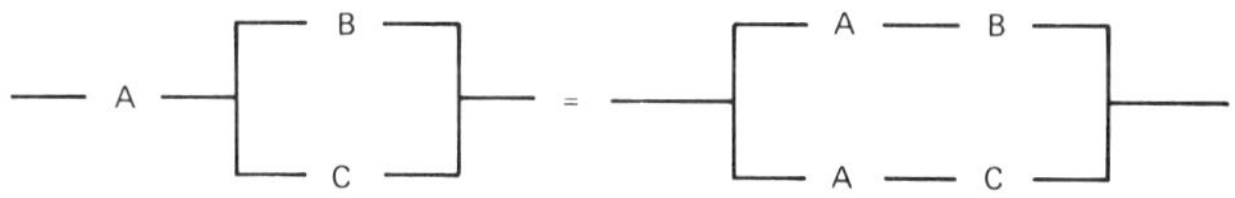

11. 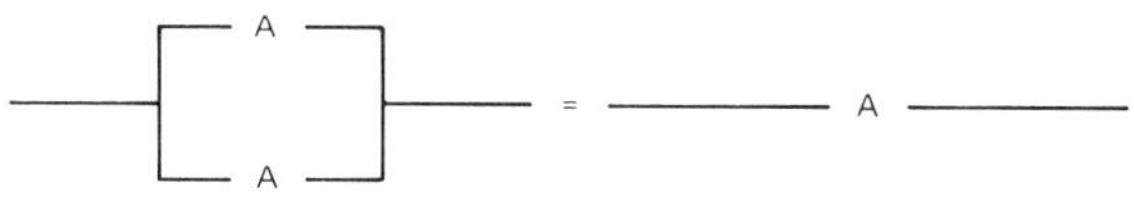

12. 

13. 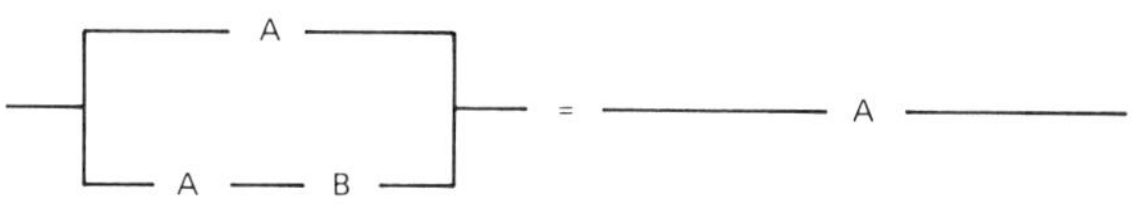

14. 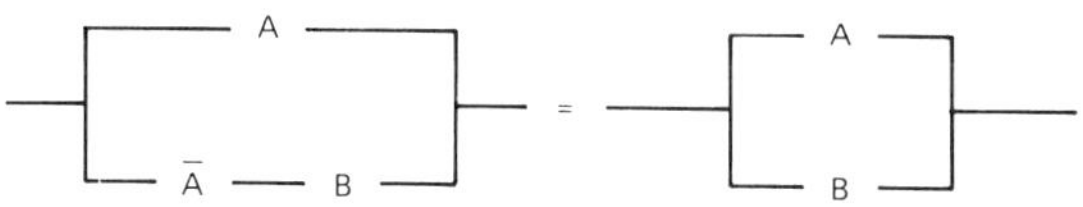

15. 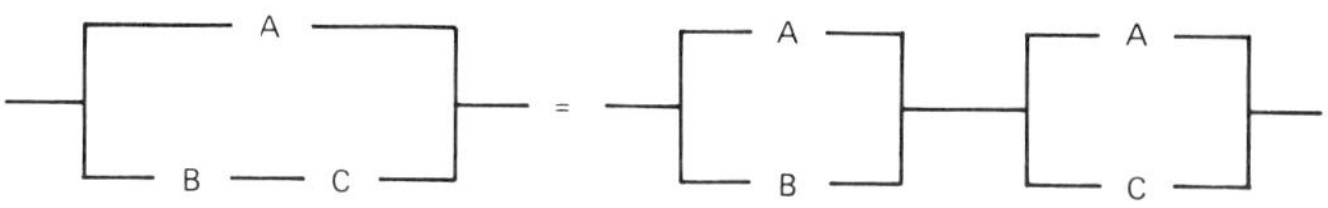

The next eight properties are unique to Boolean algebra.

Idempotent:	(Parallel)	11. $A + A = A$
	(Series)	12. $A \cdot A = A$
Special:		13. $A + AB = A$
		14. $A + \bar{A}B = A + B$
		15. $A + BC = (A + B)(A + C)$
Complementary:		16. $\bar{\bar{A}} = A$
	(Parallel)	17. $\overline{A + B} = \bar{A} \cdot \bar{B}$
	(Series)	18. $\overline{A \cdot B} = \bar{A} + \bar{B}$

The first ten properties are quite familiar and similar to the properties of any algebra. The idempotent properties are very different from properties of the algebras studied previously. If the special properties are interpreted with respect to switching circuits, the properties should make sense. These properties are used in simplifying Boolean expressions. The properties can be demonstrated in truth tables or by algebra.

$A + AB = A$

A	B	$A \cdot B$	$A + AB$
1	1	1	1
1	0	0	1
0	1	0	0
0	0	0	0

$A = A + AB$

By Algebra

$A + AB$ can be factored as $A(1 + B)$. $1 + B = 1$, so $A(1 + B) = A(1) = A$. Therefore, $A + AB = A$.

$A + \bar{A}B = A + B$

A	B	$\bar{A}$	$\bar{A} \cdot B$	$A + \bar{A}B$	$A + B$
1	1	0	0	1	1
1	0	0	0	1	1
0	1	1	1	1	1
0	0	1	0	0	0

$A + \bar{A}B = A + B$

By Algebra

$$A + \bar{A}B = (A + \bar{A})(A + B)$$
$$= (1)(A + B)$$
$$= A + B$$

$A + BC = (A + B)(A + C)$

A	B	C	BC	$A+BC$	$A+B$	$A+C$	$(A+B)(A+C)$
1	1	1	1	1	1	1	1
1	1	0	0	1	1	1	1
1	0	1	0	1	1	1	1
1	0	0	0	1	1	1	1
0	1	1	1	1	1	1	1
0	1	0	0	0	1	0	0
0	0	1	0	0	0	1	0
0	0	0	0	0	0	0	0

$A + BC = (A + B)(A + C)$

By Algebra

$(A+B)(A+C) = AA + AC + AB + BC$	FOIL method; Property 10 applied twice
$= A + AC + AB + BC$	Property 12
$= A(1 + C + B) + BC$	Property 10
$= A(1) + BC$	Property 9
$= A + BC$	Property 8

$\overline{A+B} = \bar{A} \cdot \bar{B}$ and $\overline{A \cdot B} = \bar{A} + \bar{B}$

A	B	$\bar{A}$	$\bar{B}$	$A+B$	$\overline{A+B}$	$\bar{A} \cdot \bar{B}$	AB	$\overline{AB}$	$\bar{A}+\bar{B}$
1	1	0	0	1	0	0	1	0	0
1	0	0	1	1	0	0	0	1	1
0	1	1	0	1	0	0	0	1	1
0	0	1	1	0	1	1	0	1	1

Equivalent ($\overline{A+B}$ and $\bar{A} \cdot \bar{B}$) Equivalent ($\overline{AB}$ and $\bar{A}+\bar{B}$)

These properties are the same properties addressed in Chapter 3 and called DeMorgan's rules.

All these properties should appear reasonable. However, it is important to understand the relationship involved in the properties rather than just the literal statements. Letters can be changed and the states can be changed; but the relationships are constant. It takes practice to use and understand the relationships. In the examples, the properties are those given earlier in this chapter and refer to the property being used in the simplification.

Example

Simplify $A + B + \bar{A} + C + AC$.

$A + B + \bar{A} + C + AC$	$= A + \bar{A} + B + C + AC$	Property 1
	$= 1 + B + C + AC$	Property 5
	$= 1$	Property 9

Example

Simplify $\bar{B} + BC$.

$\bar{B} + BC$	$= \bar{B} + C$	Property 14

Example

Simplify $A + A\bar{B}C + AB\bar{C}$

$A + A\bar{B}C + AB\bar{C}$	$= A(1 + \bar{B}C + B\bar{C})$	Factor
	$= A(1)$	Property 9
	$= A$	Property 8

Example

Simplify $(A + BC)(A + D) + \bar{A}B$.

$(A + BC)(A + D) + \bar{A}B$	$= A + BCD + \bar{A}B$	Property 15
	$= A + \bar{A}B + BCD$	Property 1
	$= A + B + BCD$	Property 14
	$= A + B(1 + CD)$	Factor
	$= A + B(1) = A + B$	Properties 9 and 8

EXERCISES

Simplify the following Boolean expressions. Show steps as well as answers.

1. $\bar{B} + AB$
2. $B + \bar{A}B$
3. $A(C(A + \bar{A}B) + C)$
4. $(A(B + C))\,\bar{C}D$
5. $A(D(C + AD) + B)$
6. $C(AD + BD + CD)$
7. $AB + \bar{A} + CD$

8. $A(AC + CD + C + \bar{A} + B)$
9. $A\bar{A}(B + C)$
10. $\bar{A}B + A\bar{B}$
11. $A(B + \bar{A}C + A)$
12. $\bar{A}B + (A + \bar{B})D$
13. $\overline{A + \bar{A}B}$
14. $ABC + A\bar{B}C + \bar{A}BC$
15. $AB + ABC$
16. $AB + \bar{A}BC$
17. $\overline{B + \bar{A} + C}$
18. $\overline{B\bar{A}C + \bar{A}C}$
19. $\overline{A\bar{B} + \bar{A}B}$
20. $(\bar{A} + B)(\bar{A} + C)$
21. $(A + B)(C + D)$
22. $(B + A)(B + C) + \bar{B}C$
23. $A(\overline{\bar{A}B})$
24. $A(B + AB)$
25. $A(B + C) + A + AB$

Answers

2. $$\begin{aligned} B + \bar{A}B &= B(1 + A) && 10 \\ &= B(1) && 9 \\ &= B && 8 \end{aligned}$$

4. $$\begin{aligned} (A(B + C))\,\bar{C}D &= (AB + AC)\,\bar{C}D && 10 \\ &= AB\bar{C}D + AC\bar{C}D && 10 \\ &= AB\bar{C}D + 0 && 6 \\ &= AB\bar{C}D && 7 \end{aligned}$$

6. $$\begin{aligned} C(AD + BD + CD) &= ACD + BCD + CD && 10 \\ &= CD(A + B + 1) && 10 \\ &= CD(1) && 9 \\ &= CD && 8 \end{aligned}$$

8. $$\begin{aligned} A(AC + CD + C + \bar{A} + B) &= AAC + ACD + AC + A\bar{A} + AB && 10 \\ &= AC + ACD + AC + A\bar{A} + AB && 12 \\ &= AC + ACD + A\bar{A} + AB && 11 \\ &= AC + ACD + 0 + AB && 6 \\ &= AC(1 + D) + AB && 10 \\ &= AC(1) + AB && 9 \\ &= AC + AB, \text{ or } A(C + B) && 8 \end{aligned}$$

10. $\bar{A}B + A\bar{B}$ is in simplest form.

12. $\bar{A}B + (A + \bar{B})D = \bar{A}B + D$ because $(A + \bar{B}) = \overline{\bar{A}B}$.

14. $$\begin{aligned} ABC + A\bar{B}C + \bar{A}BC &= C(AB + A\bar{B} + \bar{A}B) && 10 \\ &= C(A(B + \bar{B}) + \bar{A}B) && 10 \\ &= C(A(1) + \bar{A}B) && 5 \\ &= C(A + \bar{A}B) && 8 \\ &= C(A + B) && 14 \end{aligned}$$

16. $$\begin{aligned} AB + \bar{A}BC &= B(A + \bar{A}C) && 10 \\ &= B(A + C) && 14 \end{aligned}$$

18. $$\begin{aligned} \overline{B\bar{A}C + \bar{A}C} &= \overline{\bar{A}C(B + 1)} && 10 \\ &= \overline{\bar{A}C(1)} && 9 \\ &= \overline{\bar{A}C} && 8 \\ &= A + \bar{C} && 18 \end{aligned}$$

20. $$(\bar{A} + B)(\bar{A} + C) = \bar{A} + BC \qquad 15$$

22. $$\begin{aligned} &(B + A)(B + C) + \bar{B}C \\ &= B + AC + \bar{B}C && 15 \\ &= B + C + AC && 14 \\ &= B + C && 13 \end{aligned}$$

24. $$\begin{aligned} A(B + AB) &= AB + AAB && 10 \\ &= AB + AB && 12 \\ &= AB && 11 \end{aligned}$$

4.4 SOME USES OF BOOLEAN ALGEBRA

Recall the rules for addition from the binary number system

$$0 + 0 = 0$$

$$0 + 1 = 1$$

$$1 + 0 = 1$$

$$1 + 1 = 10$$

The parallel-circuit design does the same addition except for $1 + 1 = 0$, with 1 to carry. To show the circuit design used for the addition of binary numbers, a slightly different schematic is used. The gates stand for entire circuits and are more manageable than the prior notation.

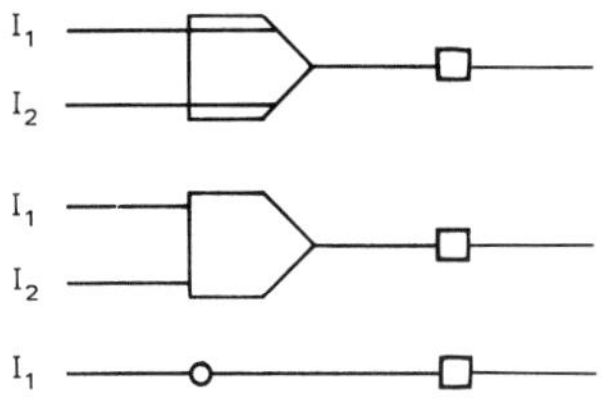

OR gate; output = $I_1 + I_2$

AND gate output = $I_1 \cdot I_2$

NOT, or complement; output = $\bar{I}_1$

The schematic that adds any two binary numbers I_1 and I_2 is as follows. Do each of the four additions above to convince yourself that the logic gates do the job.

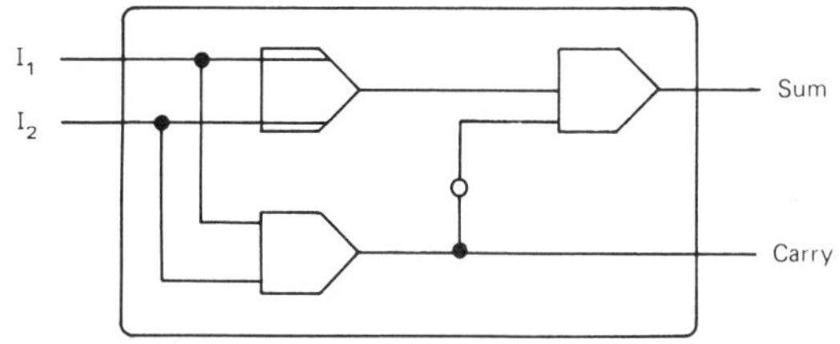

The dark dots indicate the current has taken two routes, or branched off. The empty dot indicates complement.

I_1 and I_2 are inputs and each carries a value of 0 or 1. The Boolean expression for the sum is $(I_1 + I_2)\overline{(I_1 \cdot I_2)}$ and the carry is $I_1 \cdot I_2$. If $I_1 = 1$ and $I_2 = 0$, $(1 + 0)\overline{(1 \cdot 0)}$, and the carry is $1 \cdot 0$, or $(1)(1) = 1$, with a carry of 0.

Occassionally, no circuit designs or Boolean expressions are presented; only information concerning the desired outcomes are given. Suppose a three-member committee desires an electronic device to record votes and indicate when a vote is favorable. Majority rules in this committee. Therefore, of the three variables (votes), two or three must be yea for a favorable vote. The traditional table produces all possibilities.

A	B	C	Outcome
1	1	1	1
1	1	0	1
1	0	1	1
1	0	0	0
0	1	1	1
0	1	0	0
0	0	1	0
0	0	0	0

To name this outcome: —A—B—C— $= ABC$
To name this outcome: —A—B—$\bar{C}$— $= AB\bar{C}$
To name this outcome: —A—$\bar{B}$—C— $= A\bar{B}C$
To name this outcome: —$\bar{A}$—B—C— $= \bar{A}BC$

Therefore, a majority vote is produced by ABC, $AB\bar{C}$, $A\bar{B}C$, or $\bar{A}BC$. This expression can be simplified.

$$
\begin{aligned}
ABC + AB\bar{C} + A\bar{B}C + \bar{A}BC &= AB(C + \bar{C}) + A\bar{B}C + \bar{A}BC \\
&= AB + A\bar{B}C + \bar{A}BC \\
&= A(B + \bar{B}C) + \bar{A}BC \\
&= A(B + C) + \bar{A}BC \\
&= AB + AC + \bar{A}BC \\
&= AB + C(A + \bar{A}B) \\
&= AB + C(A + B) \\
&= AB + AC + BC
\end{aligned}
$$

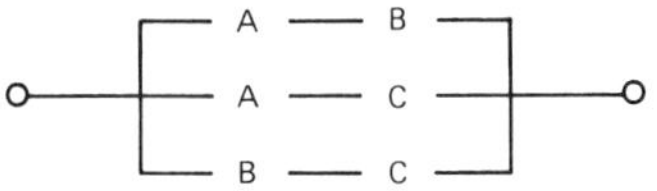

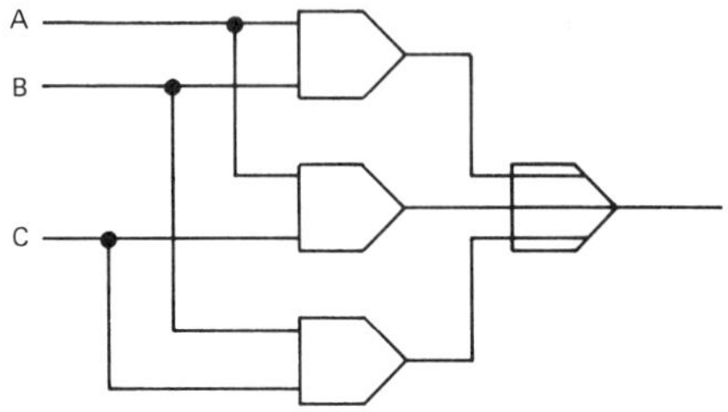

Sometimes the desired outcomes are listed in tabular form. The following table gives two desired outcomes.

A	B	C	Outcome 1	Outcome 2	Name
1	1	1	0	1	ABC
1	1	0	0	1	$AB\bar{C}$
1	0	1	1	1	$A\bar{B}C$
1	0	0	1	1	$A\bar{B}\bar{C}$
0	1	1	0	1	$\bar{A}BC$
0	1	0	0	1	$\bar{A}B\bar{C}$
0	0	1	1	0	$\bar{A}\bar{B}C$
0	0	0	1	0	$\bar{A}\bar{B}\bar{C}$

Outcome 1: $A\bar{B}C + A\bar{B}\bar{C} + \bar{A}\bar{B}C + \bar{A}\bar{B}\bar{C} = A\bar{B}(C + \bar{C}) + \bar{A}\bar{B}(C + \bar{C})$

Output: $= A\bar{B} + \bar{A}\bar{B}$

$= \bar{B}(A + A)$

$= \bar{B}$

Outcome 2 is $ABC + AB\bar{C} + A\bar{B}C + A\bar{B}\bar{C} + \bar{A}B\bar{C} + \bar{A}B\bar{C}$. This could be quite a task to simplify. In this case, it is much easier to take the nondesired outcomes $\bar{A}\bar{B}C + \bar{A}\bar{B}\bar{C}$ and invert this expression.

$$\overline{\bar{A}\bar{B}C + \bar{A}\bar{B}\bar{C}} = \overline{\bar{A}\bar{B}(C + \bar{C})}$$

$$= \overline{\bar{A}\bar{B}}$$

$$= A + B$$

If the first expression for outcome 2 is simplified, it eventually becomes $A + B$.

Example

You are wiring a two-story house and wish to install a light at the stairs that can be operated by a switch at the bottom of the stairs (A) and another switch at the top of the stairs (B). Operating a switch changes the state of the light.

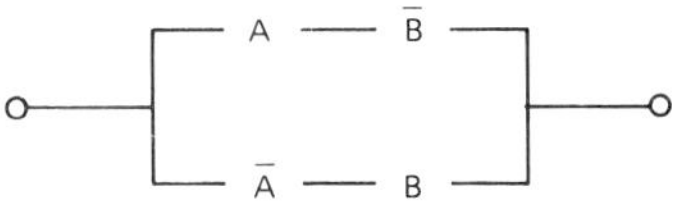

A	B	Light	A	B	Light	
Off	Off	Off	0	0	0	
On	Off	On	1	0	1	$A\bar{B}$
On	On	Off	1	1	0	
Off	On	On	0	1	1	$\bar{A}B$

Boolean expression: $A\bar{B} + \bar{A}B$

EXERCISES

1. Draw the schematic for adding two binary numbers. Let $I_1 = 1$ and $I_2 = 1$. Show the input and output at each gate and the sum and carry for this problem.
2. Draw the schematic for adding two binary numbers. Let $I_1 = 0$ and $I_2 = 1$. Show the input and output at each gate and the sum and carry for this problem.

   Given:

A	B	Outcome 3	Outcome 4	Outcome 5	Outcome 6
1	1	1	0	1	0
1	0	1	1	0	0
0	1	0	1	0	1
0	0	0	0	1	1

3. Write the Boolean expression for Outcome 3 and simplify.
4. Write the Boolean expression for Outcome 4 and simplify.
5. Write the Boolean expression for Outcome 5 and simplify.
6. Write the Boolean expression for Outcome 6 and simplify.

   Given:

A	B	C	Outcome 7	Outcome 8	Outcome 9	Outcome 10	Outcome 11	Outcome 12
1	1	1	0	1	1	0	1	1
1	1	0	1	0	0	1	1	1
1	0	1	0	0	1	0	1	0
1	0	0	0	0	0	1	0	1
0	1	1	0	0	1	0	1	0
0	1	0	0	1	0	0	1	1
0	0	1	1	0	0	0	0	1
0	0	0	0	0	0	1	1	1

7. Write the Boolean expression for Outcome 7 and simplify.
8. Write the Boolean expression for Outcome 8 and simplify.
9. Write the Boolean expression for Outcome 9 and simplify.
10. Write the Boolean expression for Outcome 10 and simplify.
11. Write the Boolean expression for Outcome 11 and simplify.
12. Write the Boolean expression for Outcome 12 and simplify.
13. A stockholder meeting has several blocks voting on each issue. Block A controls 37% of the stock, B controls 42% of the stock, C controls 9% of the stock, and D controls 12% of the stock. Write the Boolean expression for a majority vote (51% or higher), simplify, and draw the circuit diagram indicating the issue carried.
14. A stockholder meeting has several blocks voting on each issue. Block A controls 40% of the stock, B controls 42% of the stock, C controls 10% of the stock, and D controls 8%. Write the Boolean expression for a majority vote (51% or higher), simplify, and draw the circuit diagram indicating the issue carried.
15. Repeat Exercise 13, except A controls 42%, B controls 43%, and C controls 5%. Would you give C a button to push for a vote and call it a dummy?
16. Repeat Exercise 14, except A controls 41%, B controls 44%, and C controls 5%. Would you give C a button to push for a vote and call it a dummy?

Answers

2.

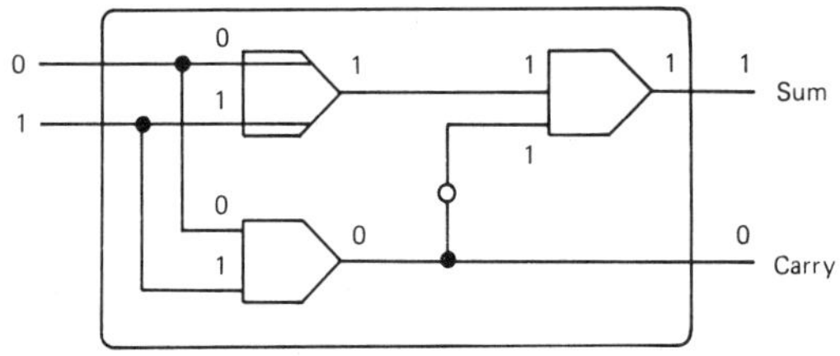

4. $A\bar{B} + \bar{A}B$ cannot be simplified.
6. $\bar{A}B + \bar{A}\bar{B}$ simplifies to $\bar{A}$.
8. $ABC + \bar{A}B\bar{C}$ cannot be simplified, but it can be written as $B(AC + \bar{A}\bar{C})$.
10. $AB\bar{C} + A\bar{B}\bar{C} + \bar{A}\bar{B}\bar{C}$ simplifies to $\bar{C}(A + \bar{B})$.
12. $(\bar{A} + B + \bar{C})(A + \bar{B} + \bar{C})$
14. $AB + BC + ACD$

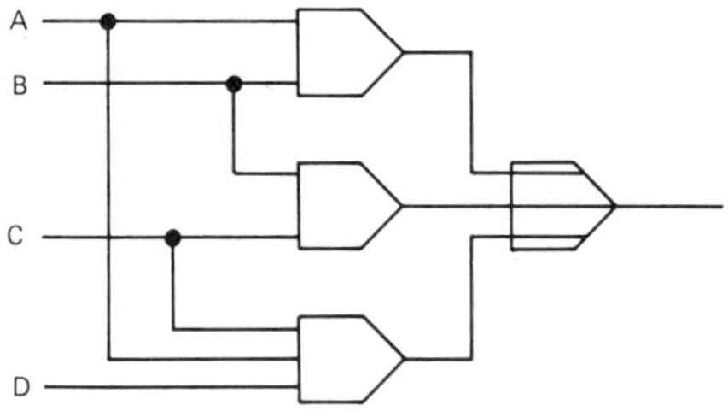

16. AB

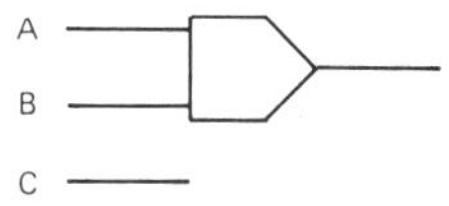

CHAPTER 4 REVIEW EXERCISES

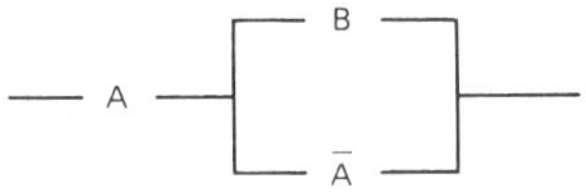

1. Find the output of the above diagram if $A = 1$ and $B = 0$.
2. Find the output of the above diagram if $A = 1$ and $B = 1$.
3. Draw a schematic for $A(B + C) + ABC$. Simplify the expression and draw the schematic for the simplified expression. Check $A = 1$, $B = 0$, and $C = 0$ to see if the outcome of each schematic is like the other.
4. Draw a schematic for $\bar{A}\bar{B}(AB + C)$. Simplify the expression and draw the schematic for the simplified expression. Check $A = 1$, $B = 0$, and $C = 1$ to see if the outcome of each schematic is like the others.

Write the Boolean expression for each of the following. Simplify. Draw the simplified circuit.

5.

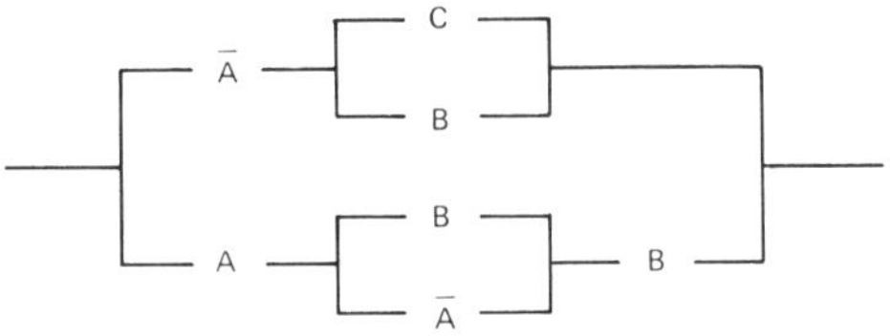

6.

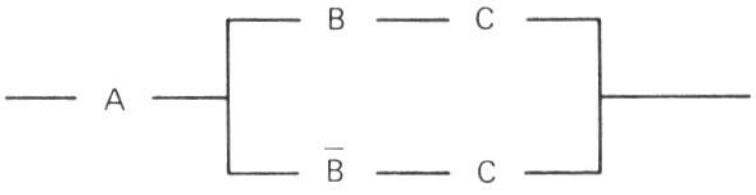

7.

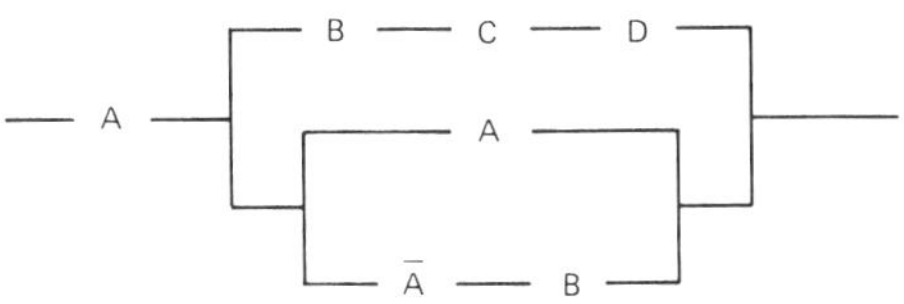

8.

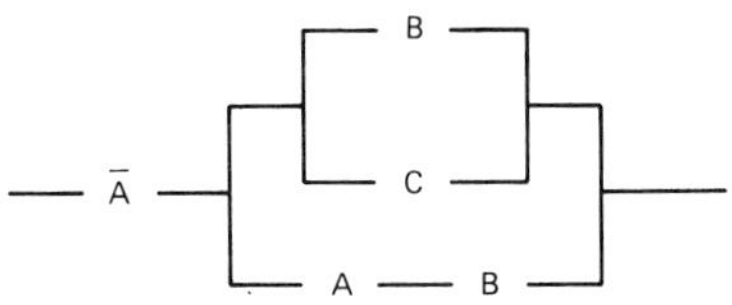

Use the properties of Boolean algebra to simplify the following:

9. $B(\overline{AB})$
10. $A(B + C) + A + AB$
11. $AB + \bar{A}$
12. $AB + \bar{A}B$
13. $(B + \bar{B})(B + C) + A\bar{B}$
14. $(B + A)(B + C) + \bar{B}C$
15. $\overline{(\bar{A} + B + CD)}$
16. $\overline{(A + \bar{A}B + C)}$

Given:

A	B	C	Outcome 17	Outcome 18
1	1	1	0	0
1	1	0	0	0
1	0	1	0	1
1	0	0	1	1
0	1	1	1	0
0	1	0	1	0
0	0	1	0	0
0	0	0	0	1

17. Write the Boolean expression for Outcome 17. Simplify.
18. Write the Boolean expression for Outcome 18. Simplify.

Chapter 4 Review Answers

2. 1
4.

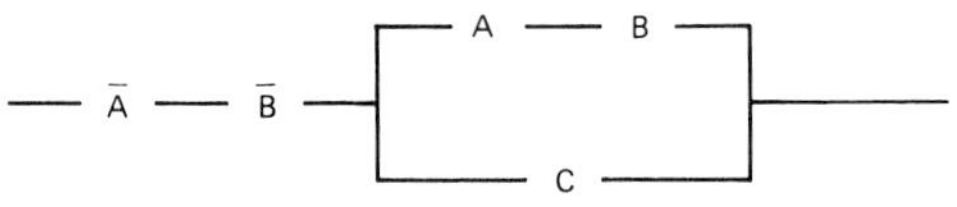

$\bar{A}\bar{B}C$; output = 0

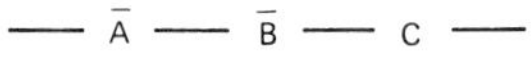

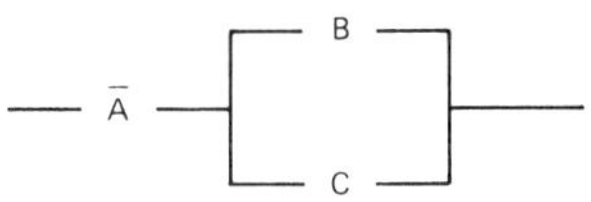

6. $A(B\bar{C} + \bar{B}C)$ cannot be simplified.
8. $\bar{A}(B + C + AB)$ simplifies to $\bar{A}(B + C)$.
10. A
12. B
14. $B + C$
16. $\bar{A}\bar{B}\bar{C}$
18. $\bar{B}(A + \bar{C})$

The Tie-Up

CHAPTER 5

5.0 INTRODUCTION

Sets, logic, and Boolean algebra have each provided notation and a set of properties. Each notation provides a vehicle to generalize properties. The specific interpretation for a particular problem does not affect the results. The outcomes are clarified by applying simplification techniques, and outcomes are logical and predictable.

The properties of binary-number arithmetic are also closely related to the properties of sets, logic, and Boolean algebra. Binary arithmetic has two bits, 0 and 1. Sets have two states—an element either belongs to a set or it does not. Logic statements are classified as true or false. Boolean expressions either have a value of 0 or 1. All these systems are called *two-state systems.*

A two-state system is the basis of a digital computer. Translation from the real world to a two-state system employs sets, logic, Boolean algebra, and binary numbers. Normally a blend of all systems is employed. It is the responsibility of the designer or programmer to select the system most appropriate. Individual preferences influence the selection also.

This chapter is designed to:

1. Review the notation of sets, logic, and Boolean algebra.
2. Point out the similarities among the systems and provide practice in translation among the systems.

3. Review the graphical representations of the systems and provide opportunities to translate among the representations.
4. Provide some applied problems involving two-state systems.

5.1 THE SIMILARITIES

The last three chapters have dealt with sets, logic, and logic circuits. These three systems have much in common. Different interpretations can be given to an expression, but the basic properties are the same for all three systems.

	Sets	Logic	Logic Circuits
Basic Component	Set	Statement	Switch
Connective	Union, $\cup$	Or, $\vee$	Parallel, +
	Intersection, $\cap$	And, $\wedge$	Series, $\cdot$
	Complement, $'$	Not, $\sim$	Inverse, $\bar{\ }$
Graphs	Venn diagram	Flowchart	Circuit
Two states	Universal set, $\mathfrak{U}$	True, T	1
	Empty set, ϕ	False, F	0
Properties			
Commutative	$A \cup B = B \cup A$	$p \vee q = q \vee p$	$A + B = B + A$
	$A \cap B = B \cap A$	$p \wedge q = q \wedge p$	$A \cdot B = B \cdot A$
Associative	$A \cup (B \cup C) =$ $(A \cup B) \cup C$	$p \vee (q \vee r) =$ $(p \vee q) \vee r$	$A + (B + C) =$ $(A + B) + C$
	$A \cap (B \cap C) =$ $(A \cap B) \cap C$	$p \wedge (q \wedge r) =$ $(p \wedge q) \wedge r$	$A(BC) = (AB)C$
Distributive	$A \cup (B \cap C) =$ $(A \cup B) \cap (A \cup C)$	$p \vee (q \wedge r) =$ $(p \vee q) \wedge (p \vee r)$	$A + BC =$ $(A + B)(A + C)$
	$A \cap (B \cup C) =$ $(A \cap B) \cup (A \cap C)$	$p \wedge (q \vee r) =$ $(p \wedge q) \vee (p \wedge r)$	$A(B + C) =$ $AB + AC$
Idempotent	$A \cup A = A$	$p \vee p = p$	$A + A = A$
	$A \cap A = A$	$p \wedge p = p$	$A \cdot A = A$
Identity	$A \cup \mathfrak{U} = \mathfrak{U}$	$p \vee T = T$	$A + 1 = 1$
	$A \cup \phi = A$	$p \vee F = p$	$A + 0 = A$
	$A \cap \mathfrak{U} = A$	$p \wedge T = p$	$A \cdot 1 = A$
	$A \cap \phi = \phi$	$p \wedge F = F$	$A \cdot 0 = 0$
Complement	$A \cup A' = \mathfrak{U}$	$p \vee \sim p = T$	$A + \bar{A} = 1$
	$A \cap A' = \phi$	$p \wedge \sim p = F$	$A \cdot \bar{A} = 0$
	$\mathfrak{U}' = \phi$	$\sim T = F$	$\bar{1} = 0$
	$\phi' = \mathfrak{U}$	$\sim F = T$	$\bar{0} = 1$
DeMorgan's rules	$(A \cup B)' = A' \cap B'$	$\sim(p \vee q) =$ $\sim p \wedge \sim q$	$\overline{A + B} = \bar{A} \cdot \bar{B}$
	$(A \cap B)' = A' \cup B'$	$\sim(p \wedge g) =$ $\sim p \vee \sim q$	$\overline{A \cdot B} = \bar{A} + \bar{B}$

In logic there were other connectives besides the three listed, but equivalences were established between these and the three basic connectives listed.

In logic and logic circuits, truth tables and output tables were useful tools to show equivalence. Sets lacked such a table, but they did utilize Venn diagrams to show two sets of elements represented by the same region of a Venn diagram. Many of the equivalences and tautologies discovered in logic and logic circuits can be translated into set notation.

Example

Logic Circuits	Logic	Sets
$AC + \bar{A}$	$(p \wedge q) \vee \sim p$	$(A \cap C) \cup A'$
$\bar{A} + AC$	$(\sim p \vee p) \wedge (\sim p \vee q)$	$(A' \cup A) \cap (A' \cup C)$
$\bar{A} + C$	$T \wedge (\sim p \vee q)$	$\mathcal{U} \cap (A' \cup C)$
	$(\sim p \vee q)$	$A' \cup C$

Example

Sets	Logic	Logic Circuits
$(B \cap C) \cup (B \cap C')$	$(p \wedge q) \vee (p \wedge \sim q)$	$BC + B\bar{C}$
$B \cap (C \cup C')$	$p \wedge (q \vee \sim q)$	$B(C + \bar{C})$
$B \cap \mathcal{U}$	$p \wedge T$	B
B	p	

Example

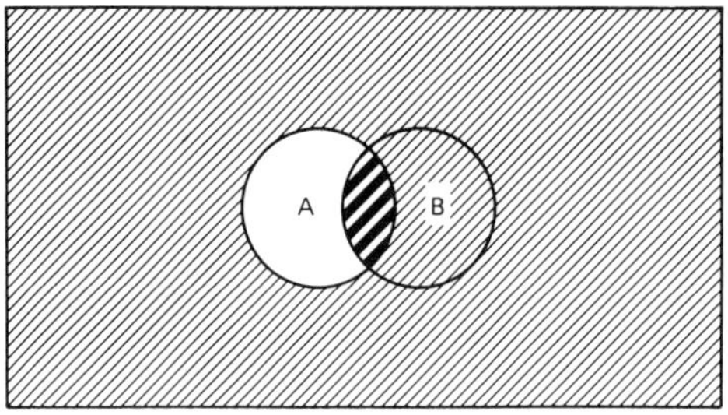

Sets $(A \cap B) \cup A' \cup B$ or

$(A \cap B')'$

$A' \cup B$

Logic $(p \wedge q) \vee \sim p \vee q$ or

$\sim(p \wedge \sim q)$

$\sim p \vee q$

Logic Circuits	$AB + \bar{A} + B$
	$\bar{A} + B + B$
	$\bar{A} + B$

p	q	$(p \wedge q) \sim P$		$((p \wedge q) \vee \sim p)$	$((p \wedge q) \vee \sim p \vee q)$
T	T	T	F	T	T
T	F	F	F	F	F
F	T	F	T	T	T
F	F	F	T	T	T

The techniques of each of the different notations yield the same result. However, the results are more mathematically obvious in the logic-circuit notation than in the other two. Some exercises will help review all three notations and allow the choice of the easiest way.

EXERCISES

Write the given examples in al! three notations, simplify in one of these notations, and write the result in all three notations.

1. (not p or r) or (q and (p or r))
2. p and (p or q)
3. P or (p and q)
4. not (p and not q) or q
5. not p and not (p or q)
6. p or (q and r) or not p
7. p and not q or r and not q
8. p and not q or q
9. (not p or q) and (not p or r)
10. p or q and (r or not s) or not p
11. If not p, then (p or q)
12. (p eor q) and not q
13. (p if and only if not q) or not p
14. not (p and (not p or r))
15. not ((p and not q) or q)

Answers

2. $P(p + q) = p + pq = p$
   $p \wedge (p \vee q) = p$
   $p \cap (p \cup q) = p$

4. $\overline{p\bar{q}} + q = \bar{p} + q + q = \bar{p} + q$

$\sim(p \wedge \sim q) \vee q = \sim p \vee q$

$(p \cap q')' \cup q = p' \cup q$

6. $p + (qr) + \bar{p} = 1$

$p \vee (q \wedge r) \vee \sim p = T$

$p \cup (q \cap r) \cup p' = \mathfrak{U}$

8. $p\bar{q} + q = p + q$

$(p \wedge \sim q) \vee q = p \vee q$

$(p \cap q') \cup q = p \cup q$

10. $p + q(r + \bar{s}) + \bar{p} = 1$

$p \vee q \wedge (r \vee \sim s) \vee \sim p = T$

$p \cup q \cap (r \cup s') \cup p' = \mathfrak{U}$

12. $(p\bar{q} + \bar{p}q)\,\bar{q} = p\bar{q}$

$((p \wedge \sim q) \vee (\sim p \wedge q)) \wedge \sim q = p \wedge \sim q$

$((p \cap q') \cup (p' \cap q)) \cap q' = p \cap q'$

14. $\bar{p} + p\bar{r} = \bar{p} + \bar{r}$

$\sim p \vee (p \wedge \sim r) = \sim p \vee \sim r$

$p' \cup (p \cap r') = p' \cup r'$

5.2 THE VERBAL PROBLEM

There is no such thing as a mathematics problem outside of the classroom. There are problems in the real world that require solutions. Mathematics is a useful tool to determine solutions. Mathematics is not the solution. Problems are either stated verbally or written. The solver must translate the problem into notation that clarifies the situation. Simplification techniques are applied to clarify the question. The solution of the problem is usually attained by applying principles of such areas as mathematics, economics, and psychology. Sets, logic, Boolean algebra, and binary arithmetic are all tools useful in problem solving.

Often a problem is solved by translating the information into notation of sets, logic, or Boolean algebra. The answer should always be translated back to the language of the question. If the question asks how old the daughter is and the solution is some mathematics with a statement $x = 18$, it requires interpretation of this result to answer the original question.

Example

Qualifications for a programmer were listed as follows:

1. A 4-year degree, 2 or more years of computer-related work, and at least 25 years of age.
2. No degree, 5 or more years of computer-related work, and at least 25 years of age.
3. No degree, but military training with at least 5 years of computer-related work.

These qualifications can be symbolized as:

$$(D \text{ and } T \text{ and } Y) \quad \text{or} \quad (D' \text{ and } F \text{ and } Y) \quad \text{or}$$

$$(D' \text{ and } M \text{ and } F) = DTY + D'F(Y + M)$$

Qualification requirements are a 4-year degree with 2 years of experience and 25 years of age or older, or no degree but 5 years of computer experience and either military training or 25 years of age.

EXERCISES

Use whatever notation seems most appropriate to describe the group. Simplify the expression and translate the solution back to the vocabulary of the original problem.

1. Blue-eyed female accountants or blue-eyed female receptionists and male accountants.
2. Yellow four-door convertibles or nonyellow four-door convertibles.
3. Nonwhite female drivers or nonwhite male drivers.
4. Customers who earn more than $35,000 and live on Nob Hill or customers who earn $35,000 or less and do not live on Nob Hill.
5. Houses with fireplaces, dishwashers, and air conditioners or apartments with fireplaces, dishwashers, and air conditioners.
6. Locations where the temperature is moderate, the humidity is low, and rainfall is average or places where the temperature is not moderate or the humidity is not low or the rainfall is not average.
7. Brown leather shoes with laces or brown nonleather shoes with laces.
8. White unemployed individuals under 25 or nonwhite unemployed individuals.
9. Non-interest-bearing 1-year debentures or interest-bearing debentures.
10. Licensed drivers with three or more points or unlicensed drivers.
11. Everything except large Valencia oranges or green bananas.
12. Everyone except white female teenagers.
13. All migratory birds that are long-billed or not long-billed.
14. No service charge only if you maintain a balance of $200 and write not more than ten checks.
15. Perishable dairy products and nondairy products are to be removed.
16. Purchase nonfattening, perishable, high-calorie foods.
17. Metric and nonmetric measure.
18. Metric or nonmetric measure.
19. Steel-belted tires or non-steel-belted white wall tires.
20. Smokers and nonsmokers are welcome.
21. Decaffinated instant coffee or caffinated chicory-flavored instant coffee.
22. Friday the thirteenth or Friday.
23. It is not the case that the suspect was a blue-eyed male teenager.

24. Single male drivers under 25 or married male drivers over 25.
25. First-in-first-out or first-in-last-out.

Answers

2. Four-door convertibles.
4. Customers who either earn over $35,000 and live on Nob Hill or do not earn over $35,000 and do not live on Nob Hill.
6. Anywhere
8. Unemployed who are nonwhite or under 25.
10. Unlicensed drivers or drivers with three or more points.
12. Nonwhite or males or nonteenager.
14. A service charge or maintain a balance of $200 and write no more than ten checks.
16. Purchase no food (if nonfattening and high-calorie are opposites!)
18. All measure
20. No one welcome
22. Friday
24. Cannot be simplified.

5.3 TRANSLATION FROM VENN DIAGRAMS

Pictorial representations of set notation are called *Venn diagrams.* Often information can be clarified by the use of Venn diagrams. Translation from Venn diagrams to sets is cumbersome and the resulting expression is often "wordy." Using logic circuit notation, the Venn diagram can be translated and simplified with greater accuracy and ease.

In Chapter 2, areas within Venn diagrams were identified and named. These same areas can be named in logic circuit notation.

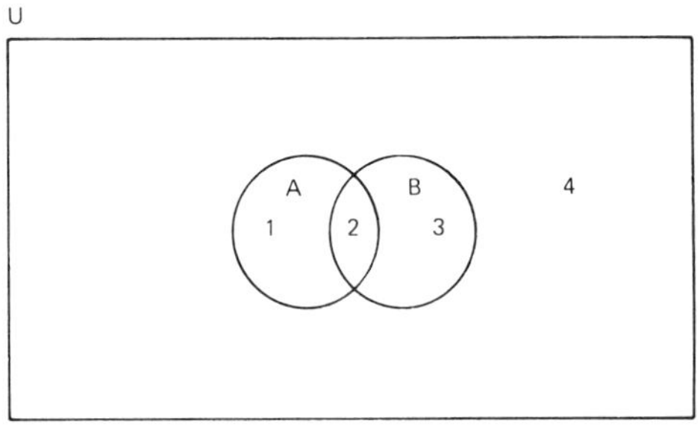

Area 1 is $A\bar{B}$
Area 2 is AB
Area 3 is $\bar{A}B$
Area 4 is $\bar{A}\bar{B}$

Likewise, in the three subset diagrams, the eight areas can be identified.

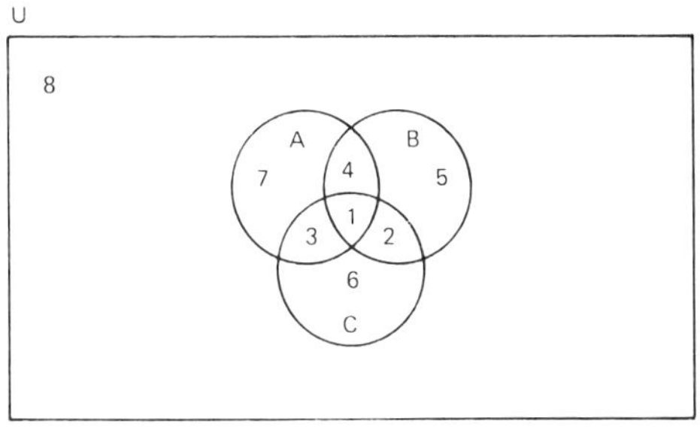

Area 1 is ABC
Area 2 is $\bar{A}BC$
Area 3 is $A\bar{B}C$
Area 4 is $AB\bar{C}$
Area 5 is $\bar{A}B\bar{C}$
Area 6 is $\bar{A}\bar{B}C$
Area 7 is $A\bar{B}\bar{C}$
Area 8 is $\bar{A}\bar{B}\bar{C}$

Using this notation, subset A would become:

$$7 + 4 + 3 + 1 = A\bar{B}\bar{C} + AB\bar{C} + A\bar{B}C + ABC$$

because there are four distinct areas making up subset A. Using the properties of Boolean algebra, this expression becomes:

$$A\bar{B}\bar{C} + AB\bar{C} = A\bar{C} \qquad A\bar{B}C + ABC = AC \qquad A\bar{C} + AC = A$$

If it is obvious that the entire subset is included, the subset would simply be named rather than going through the simplification process. However, the simplification works and takes away the need to recognize an area; naming each distinct part and simplifying gives the correct name. If an area to be named is adjacent to another area to be named, some simplification is possible. If the areas do not share any common boundary, the two names remain distinct.

Example

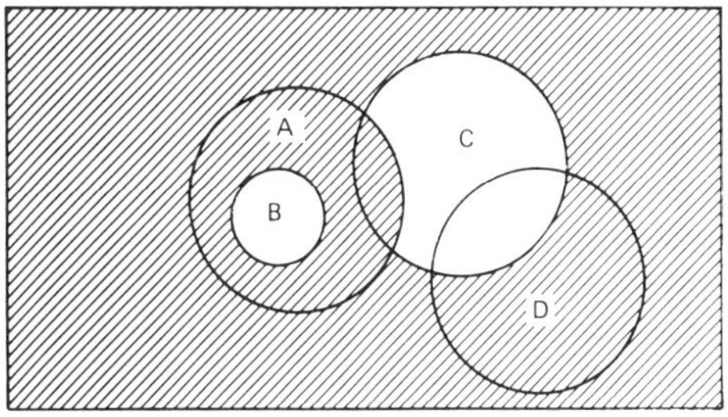

A is the set of all social security recipients

B is the set of dependent children receiving social security benefits

C is the set of people over 65 years of age

D is the set of federal employees

The shaded area can be named $A\bar{B} + \bar{C}D + \bar{A}\bar{B}\bar{C}\bar{D}$.

$$A\bar{B} + \bar{C}(D + \bar{A}\bar{B}\bar{D}) = AB + \bar{C}(D + \bar{A}\bar{B})$$
$$= A\bar{B} + \bar{C}D + \bar{A}\bar{B}\bar{C} = \bar{B}(A + \bar{A}\bar{C}) + \bar{C}D$$
$$= \bar{B}(A + \bar{C}) + \bar{C}D$$

All people not 65 years of age that are federal employees or social security recipients who are not dependent children or people not over 65 years of age that are not dependent children recipients.

Example

Represent active male military personnel, women under the age of 35, or military women, active or inactive, over the age of 35.

Let

$\mathcal{U}$ represent U.S. citizen

A represent active and inactive military personnel

B represent inactive military personnel

C represent female citizens

D represent citizens under the age of 35

The Boolean expression to represent this Venn diagram is:

$$A\bar{B} + CD + CA$$

$$A\bar{B} + C(D + A)$$

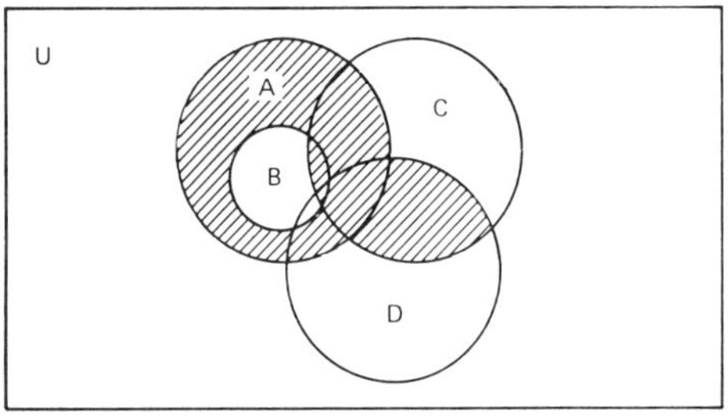

This is active military or women who are under 35 or in the military.

EXERCISES

Name each section of the shaded area, combine, simplify, and express in set notation.

1.

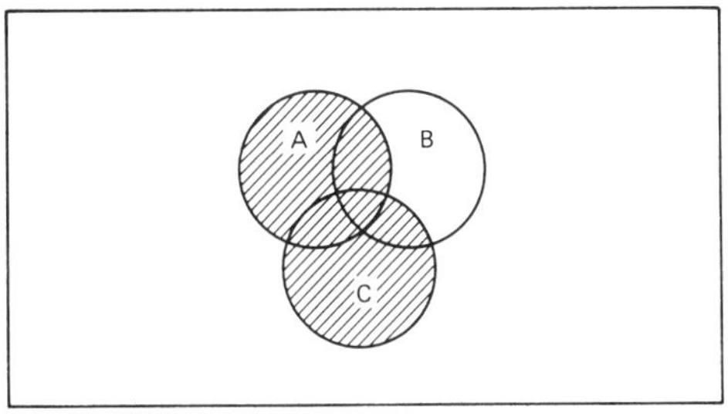

2.

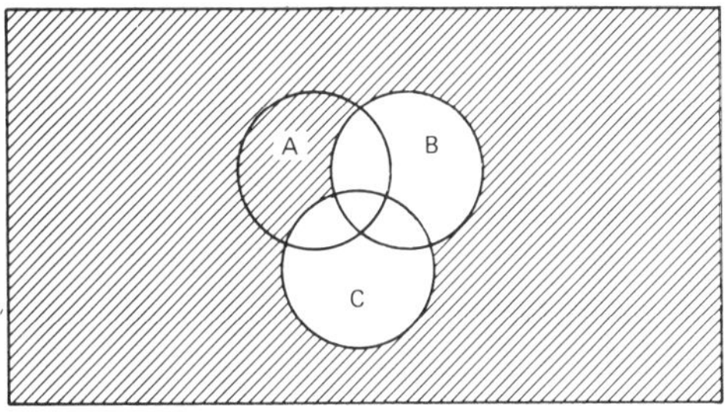

3.

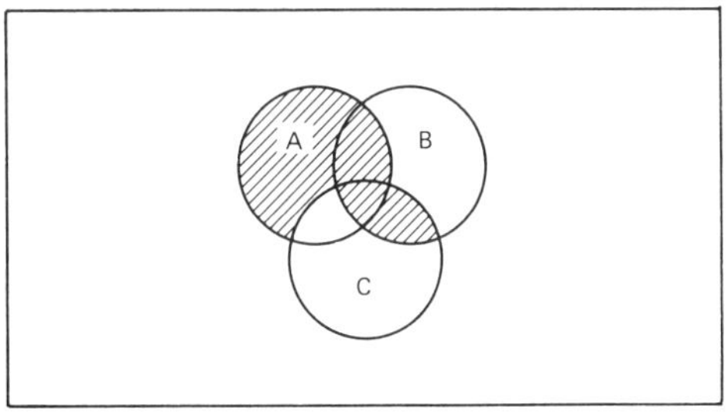

4.

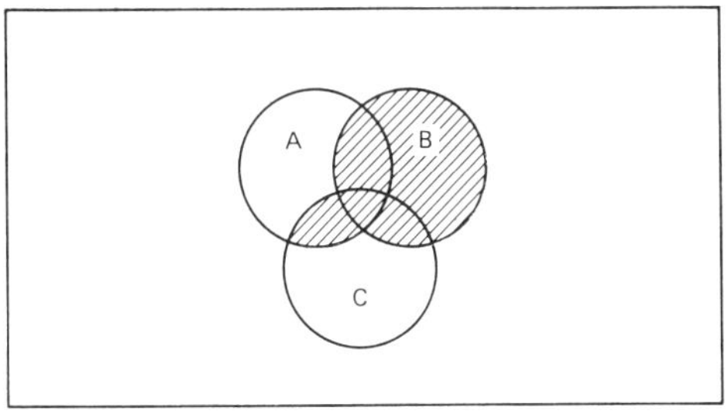

5.

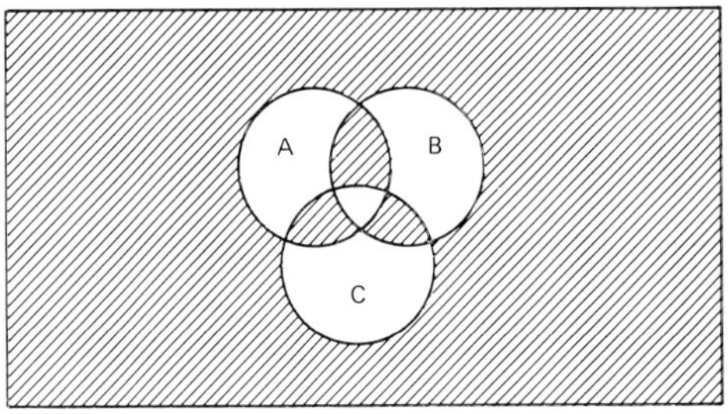

6.

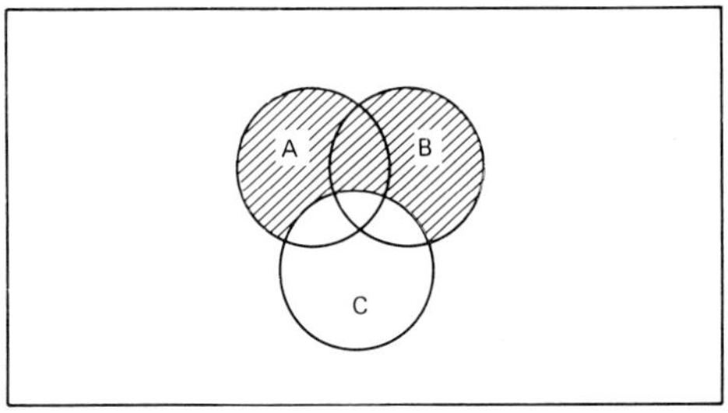

7.

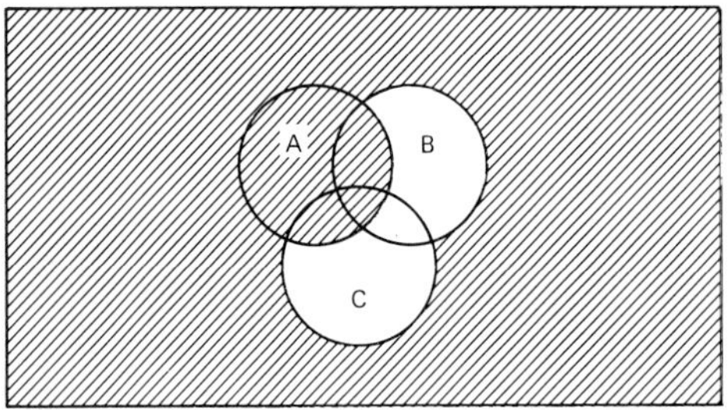

8.

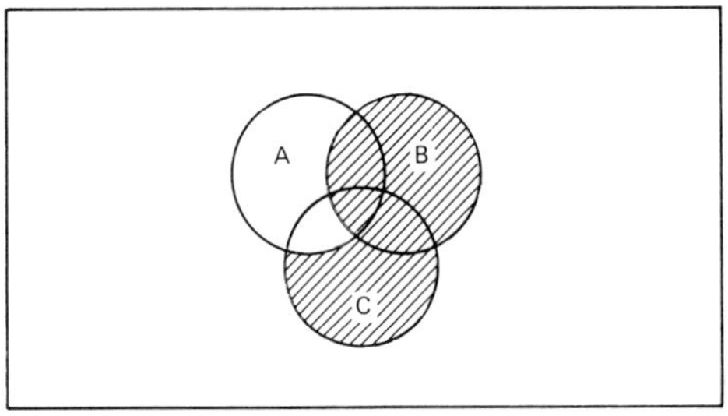

9.

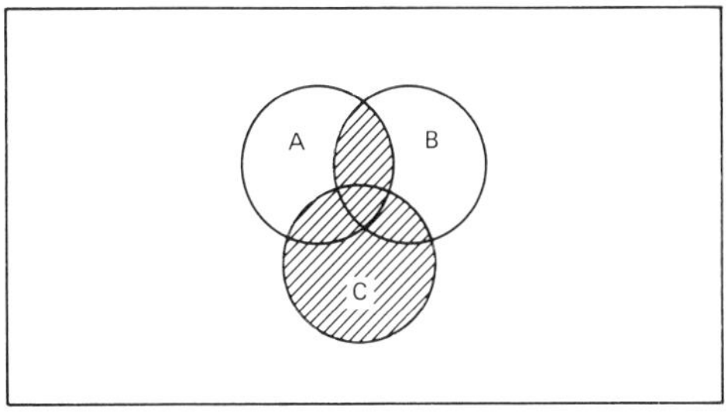

10.

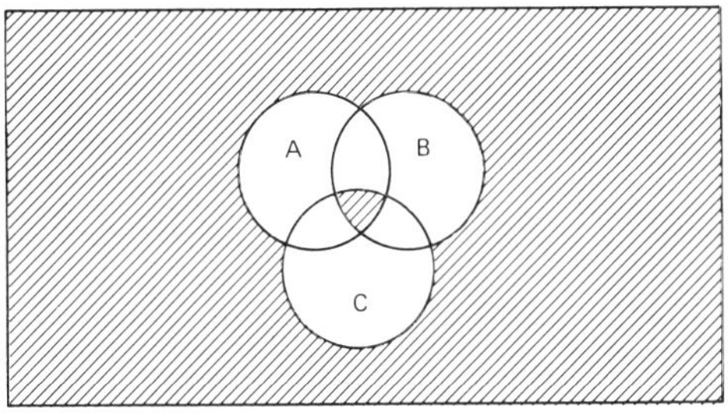

Draw a Venn diagram of the described relationship, write the Boolean expression for the relationship, simplify, and verbalize the simplified relationship.

11. Recall GM autos made in 1982 with cruise control and power brakes or other-year models with power brakes.
12. Patients who have had X-ray treatments in the past 5 years and developed thyroid problems or patients who have not had X-ray treatments but who live in Worldville and have thyroid problems.
13. All dogs that do not bark but whimper, dogs that bite, or dogs who do not whimper, bark, or bite.
14. Layoff notices went to all employees with less than 2 years service, employees with over 2 years service who worked in finishing, or weavers with more than 2 years of service but with excessive absences.

Answers

2. $A\bar{B}\bar{C} + \bar{A}\bar{B}\bar{C} = \bar{B}\bar{C} = B' \cap C'$
4. $\bar{A}B\bar{C} + AB\bar{C} + ABC + A\bar{B}C = (B \cap C') \cup (A \cap C); B\bar{C} + AC$
6. $A\bar{B}\bar{C} + \bar{A}B\bar{C} + AB\bar{C} = \bar{C}(B + A) = C' \cap (B \cup A)$
8. $\bar{A}B\bar{C} + \bar{A}\bar{B}C + ABC + AB\bar{C} + \bar{A}BC = B + AC = B \cup (A' \cap C)$
10. $ABC + \bar{A}\bar{B}\bar{C} = (A \cap B \cap C) \cup (A' \cap B' \cap C')$

12. 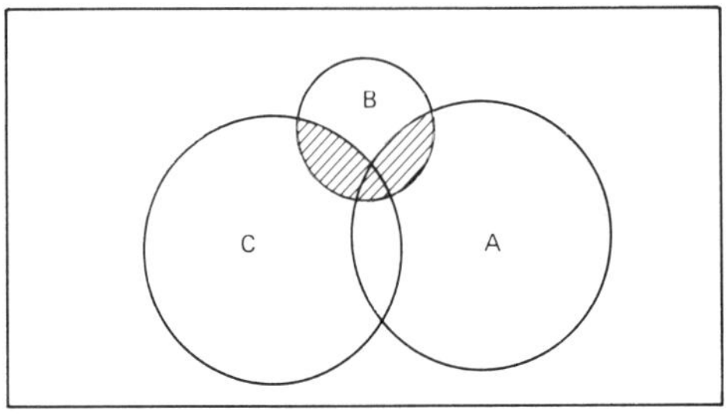

A = patients treated with X-ray
B = patients with thyroid problems
C = patients who live in Worldville
$AB\bar{C} + \bar{A}BC + ABC = B(A + C)$

Thyroid patients who have had X-ray treatment or live in Worldville.

14. 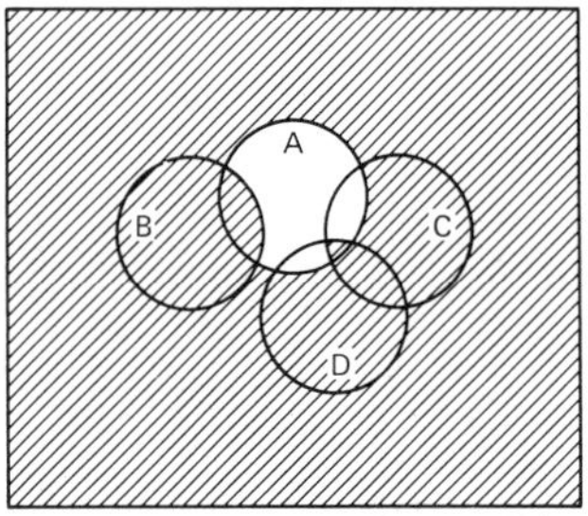

A = over two years service
B = finishing dept workers
C = weavers
D = excessive absences

Layoffs for employees with less than 2 years service or finishing department workers or weavers with excessive absences.

5.4 FLOWCHARTING

Flowcharts are the visual representations of logic. When a truth table in logic results in a true value, the flowchart representing the logical statement flows from top to bottom, indicating the conditions are met.

It is desirable to streamline a flowchart as much as possible because decisions in a flowchart represent time in a computer program.

Guidelines for Streamlining a Flowchart

1. Connectives other than NOT, AND, and OR must be replaced by some combination of these three.
2. Parentheses must be cleared.
3. In general, the number of decision boxes should not exceed the number of statements. This is not always possible. The connective *exclusive or* requires three decision boxes for two statements.
4. In an AND decision-box arrangement, the first decision box should represent the most-restrictive condition. If it is not met, no other conditions are checked.
5. In an OR decision box arrangement, the first decision box should repre-

sent the least-restrictive condition. If the decision is met, no other decision is necessary.

Example

A list is needed of employees who have over 25 years of service and are not on leave of absence.

Far more employees will be eliminated from the condition "over 25 years of service" than "not on leave of absence." Therefore, look at the number of years of service first. With 1000 employees, the order of decision boxes can make a big difference in the computer time required for this printout.

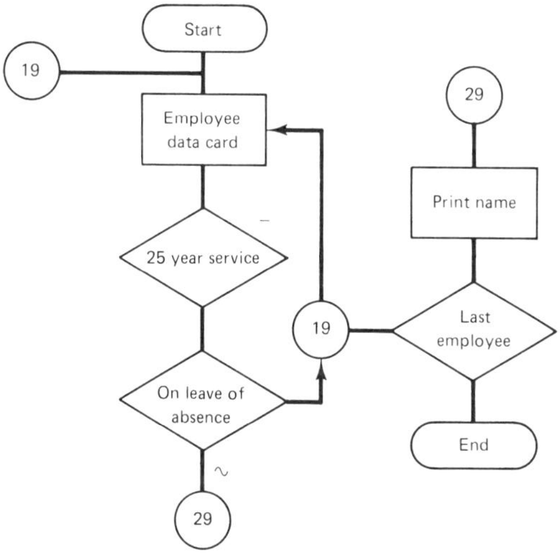

Example

Simplify and arrange the decision boxes that will represent NOT (p EOR NOT q) when it is true.

1. Replace EOR using the rules of logic.

$$p \text{ EOR NOT } q = \text{NOT } p \text{ AND NOT } q \text{ OR } p \text{ AND } q$$

2. NOT ((NOT p AND NOT q) OR (p AND q))

$\overline{\bar{p}\bar{q} + pq}$	Write this in logic circuit notation.
$= (p + q)(\bar{p} + \bar{q})$	Apply the NOT.
$= p\bar{p} + p\bar{q} + \bar{p}q + q\bar{q}$	Distributive property
$= p\bar{q} + \bar{p}q$	Simplify.

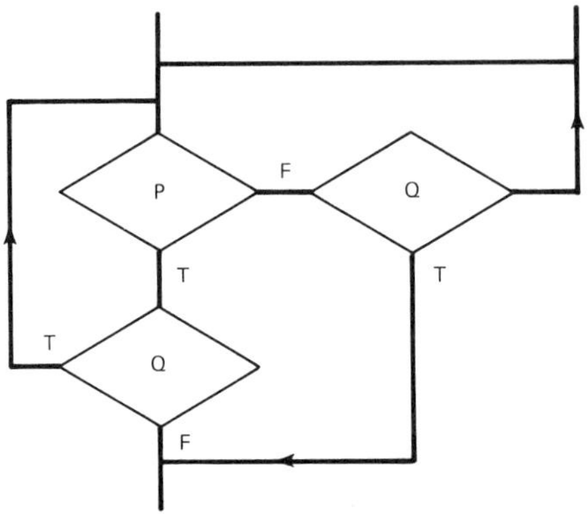

This is an example where it is necessary to have more decision boxes (three) than statements (two). The simplification utilizes both logic (replacement for EOR) and Boolean algebra (complement and product). In an example such as this, no priority on order can be shown because it is not known what statements P and Q represent.

EXERCISES

In the following, write each in terms of AND, OR, or NOT, clear the parentheses, simplify, and flowchart the decision boxes.

1. NOT (p OR q) OR q
2. p OR NOT (p AND NOT q)
3. (p OR q) $\longrightarrow$ NOT q
4. NOT q $\longrightarrow$ (p AND q)
5. p EOR q OR NOT p
6. NOT p EOR (q OR p)
7. NOT p $\longleftrightarrow$ (q AND p)
8. $q \rightarrow$ (p OR q)
9. NOT (p(p OR q))
10. NOT (q(p AND q))

In the following, identify each statement by a letter, write the logical form of the statement, simplify if possible, arrange the decision boxes in the most efficient manner.

11. Taxpayers who earn more than \$10,000, who itemize deductions, and sold a residence in the preceding year.
12. Students whose legal residence is in the state where the college is located, who major in medieval literature, and who maintain a 3.0 GPR.
13. Citizens of the United States who shop in grocery stores, take public transportation to work, or collect old coins.
14. Trees that have green leaves, smooth bark, OR sap that makes maple syrup.
15. If the president is reelected and serves out the four-year term, the year is 1989.
16. The crop will be bountiful if the rainfall is adequate or the irrigation system is functioning.

Answers

2. p OR NOT p OR $q = p + \bar{p} + q = 1$; always true, no decision boxes.
4. q OR (p AND q) $= q + pq = q$

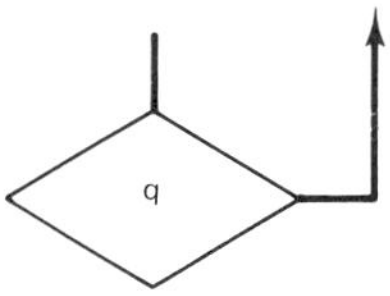

6. (p AND q) OR (NOT p AND NOT q) OR $p = pq + \bar{p}\bar{q} + p = p + \bar{q}$

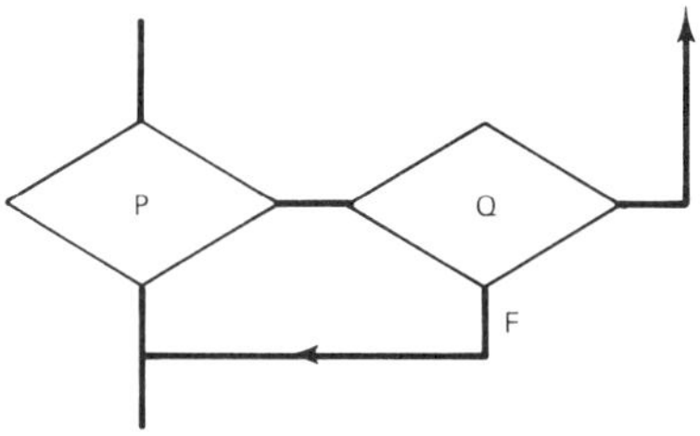

8. NOT Q OR P OR $Q = \bar{Q} + P + Q = 1$; always true, no decision boxes.
10. $\overline{(pqq)} = \overline{(pq)} = \bar{p} + \bar{q}$

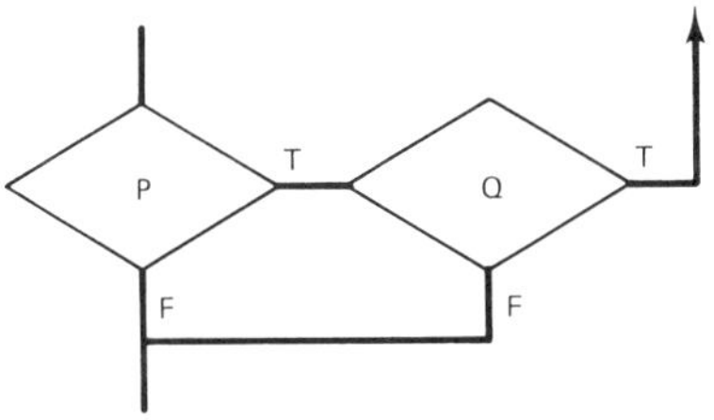

12. p: students whose legal residence is the state where the college is located
    q: students who major in medieval literature
    r: students who maintain a 3.0 GPR

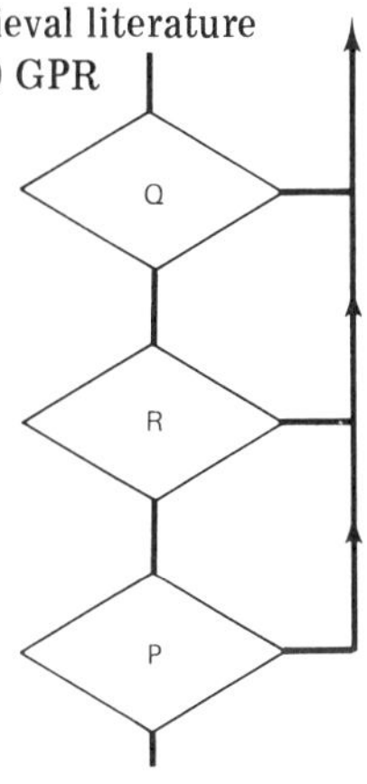

14. p: trees with green leaves
    q: trees with smooth bark
    r: trees with sap that makes maple syrup

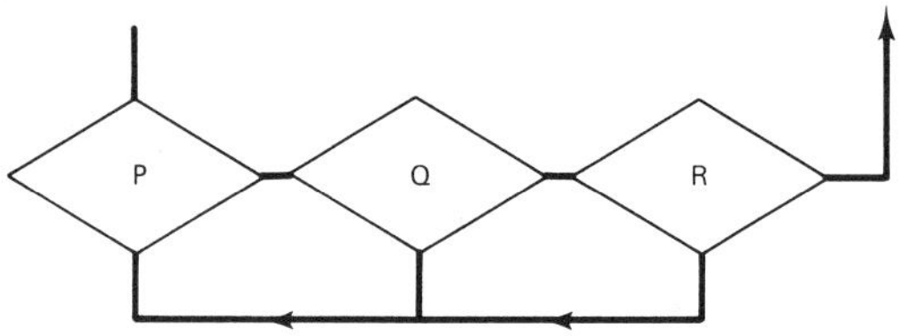

16. p: bountiful crop
    q: adequate rainfall
    r: functioning irrigation system

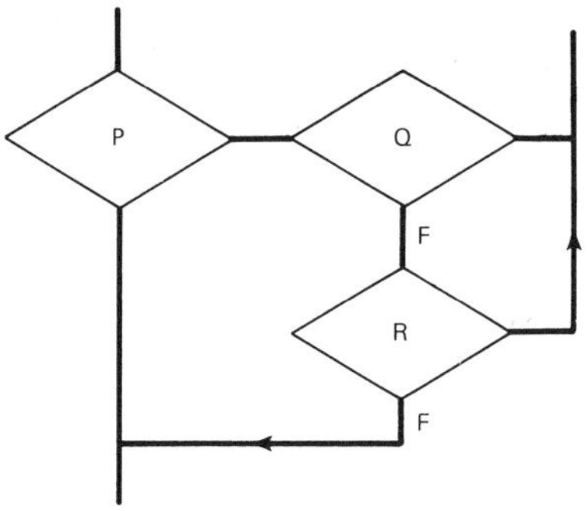

5.5 SOME ADDITIONAL USES

Information is frequently stored in a computer and called up as needed. The information is in the yes-no variety and is stored as 1 or 0. The location of each 1 or 0 in a several-bit number gives the meaning to the 1 or 0.

Example

Suppose a community college has many applicants for part-time teachers. To organize the information and facilitate finding an applicant, the information about applicants is stored on a computer. The qualifications that are decided upon are:

Master's degree
Bachelor's degree
Five years or more in field
Available during day
Available in evening

Biology
Math
English
History
Business

There are ten different qualifications stored; therefore, a 10-bit storage area is used. Suppose an applicant has a master's degree, 10 years of experience in the field, is available in the evening, and is qualified to teach English and History. This information is stored as 1110100110. The second 1, for a bachelor's degree, is assumed because if a person has a master's degree, it is usually preceded by a bachelor's degree.

A mathematics instructor is needed to teach a course during the day. The necessary qualifications are a master's degree, 5 or more years experience in the field, day class, and math course. The code put in the computer is 1111001000. All applicants are checked against this code using the AND conjunction, bit by bit.

Qualifications needed:	1111001000
Applicant:	1110100110
AND:	1110000000

This shows the applicant had the first three qualifications required but lacked the remainder needed.

Sometimes it is more beneficial to find the qualifications that are lacking. The EOR connective is used in this situation.

Qualifications needed:	1111001000
Applicant:	1110100110
EOR	0001101110

The qualifications that failed to match are teaching during the day or night and the subjects math, English, and history.

EXERCISES

1. Suppose you have decided to start up a computerized partner match. Choose eight characteristics you would match. Enter your qualifications in each category. Enter your "dream" partner's qualifications. How many of the dream characteristics do you match? What are the characteristics that do not fit?
2. A national real estate office classifies property and puts the information on the computer. Select the ten qualifications you would enter for a piece of real estate. Classify your property (or your parents' property) in each of these categories. If you are buy-

ing a piece of real estate, qualify what you wish to have. How many qualifications does your present property have and what are the characteristics that do not fit?

3. Set up a program for screening applicants for entrance to XYZ University. Qualifications are a high school diploma, 800 or higher on SAT, 3 years math, top one-third of the graduating class, a foreign language, income (of parents) over $15,000 per year, resident of state, and 18 years of age. Provide a printout to the potential student to show where the discrepancies in qualifications exist. The applicant is a 19-year-old state resident whose father's income is $20,000. She is a high school graduate with SAT scores of 750, speaks 2 foreign languages, took general math, and graduated in the top half of her class.
4. Set up a program to determine possible graduates. Qualifications for graduation are 9 credits of English composition, 9 credits of speech or literature, 18 credits of social science, 24 credits of mathematics, 24 credits of science, 18 credits of humanities, GPR of over 2.0, and graduation fee of $15 paid. Determine the requirements John Doe possesses if he has 9 credits of English, 6 credits of speech and literature, 21 credits of social science, 22 credits of mathematics, 30 credits of science, 18 credits of humanities, and a GPR of 2.67.

CHAPTER 5 REVIEW EXERCISES

1. A customer desires a six-cylinder car with automatic transmission and power steering or manual transmission only if the car is not six-cylinder. Draw a Venn diagram representing the desired car. Write a Boolean expression representing the desired car. Simplify the Boolean expression. Design the flowchart for the desired car.
2. A doctor prescribed $\frac{1}{2}$ hour exercise, no medication, and adequate rest; medication and adequate rest; or exercise, medication, and rest. Draw a Venn diagram representing the prescription. Write the Boolean expression representing the prescription. Simplify the Boolean expression. Design the decision boxes of the flowchart to determine prescription.
3. It is not true that if you smoke you will get cancer. Under what conditions is this a true statement?
4. It is not a fact that I am a liar and a cheat. When is this a true statement?
5. Qualifications for pilots in the Air Force were set as follows: college graduates or IQ above 130 and 3 years military experience. The senate committee wanted to set the qualifications of college graduate, IQ above 130, and 3 years military experience. Which are the most restrictive qualifications? What is the intersection of the two sets of qualifications? What is the union of the two sets of qualifications?
6. A leading economist made the following statement: If unemployment decreases, interest rates will increase. The recession will bottom out or interest rates will decrease. If the recession does not bottom out, unemployment will increase. What did the economist say?

Chapter 5 Review Answers

2.

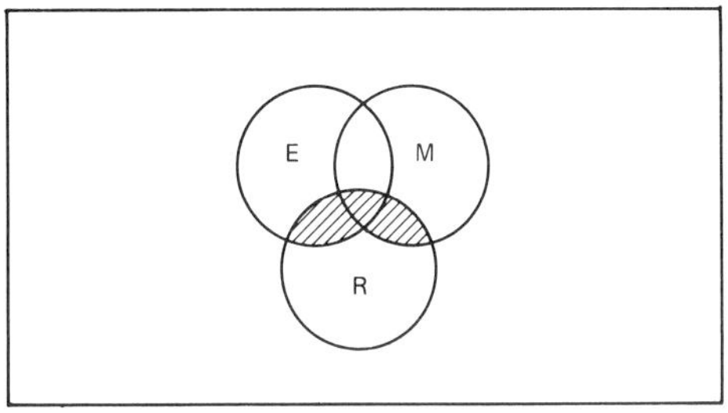

$E\bar{M}R + \bar{E}MR + EMR$

$R(E + M)$

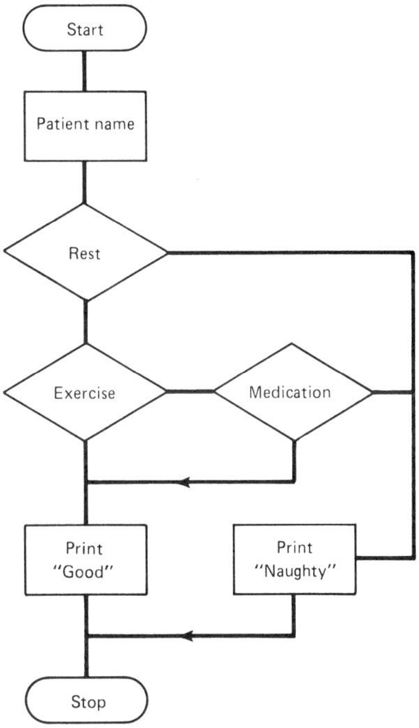

4. True under all conditions except when "You are a liar" is true and "You are a cheat" is true.
6. Unemployment increases and the recession bottoms out or interest rates increase and the recession bottoms out or interest rates decrease and unemployment decreases.

Algebra

CHAPTER 6

6.0 INTRODUCTION

This chapter assumes some knowledge of algebra. The presentation includes a review of basic algebraic skills integrated with computer requirements of algebraic notation.

The algebraic skills developed are directed toward the solution of linear equations and inequalities, quadratic equations, and higher-degree equations. Integration of flowcharting and iteration incorporates computer techniques into the solution of equations.

The algebraic skills practiced in solving numerical equations are used in solving literal equations and formulas. The solutions are flowcharted to indicate the proper sequence and necessary procedures for a required output.

This chapter is designed to:

1. Provide the mathematical vocabulary to communicate and function in the computer arena.
2. Provide a review of algebraic skills, with special emphasis on methods and notation utilized in programming.
3. Provide the skills and concepts needed to solve linear, quadratic, and higher-degree equations.

6.1 ALGEBRAIC EXPRESSIONS

Some basic definitions are given to insure a clear understanding of the materials that follow in the chapter.

Algebraic expression. An *algebraic expression* is a collection of numbers and symbols related to one or more arithmetic operations. The letters are called *variables* because their values can vary from situation to situation.

Factor. Factors are the quantities involved in a product.

Term. *Terms* are the quantities involved in an addition or subtraction.

Example

$3xy + z$ is an algebraic expression.

3, x, y, $3x$, $3y$, and xy are all factors of $3xy$.

$3xy$ and z are terms of the algebraic expression.

Coefficient. A *coefficient* is the factor of a given letter or group of factors. Naming the coefficient in the expression $3xy$ cannot be done, but the coefficient of y can be given; the answer is $3x$.

An algebraic expression usually involves more than one arithmetic operation. A hierarchy of operations in algebra has been established:

1. Exponents and radicals
2. Multiplication and division
3. Addition and subtraction

If there is more than one multiplication or division within an expression, the operations are performed as you move from left to right through the expression.

Parentheses or grouping symbols can change the order of operations; all operations within parentheses must be performed before the hierarchy of operations takes over.

Evaluating an Expression

To *evaluate an expression* means to substitute particular values for the variables in an algebraic expression, perform the indicated operations as outlined in the hierarchy of operations, and find a numerical value for the expression.

To evaluate the algebraic expression $2x - 3y$, a value must be given to each of the variables x and y.

If $x = 4$ and $y = -2$, the letters x and y are replaced by the corresponding given values. It is useful to use parentheses for each value because the algebraic way of writing two letters is different from the arithmetic meaning;

$$2x - 3y = 2(4) - 3(-2) \qquad \text{for} \quad x = 4, y = -2$$

The hierarchy of operations dictates the multiplications be performed first.

$$2(4) - 3(-2) = (8) - (-6)$$

Now the subtraction is performed, giving a value of 14 to the expression.

Careful attention must be given to the rules of signed numbers in the evaluation of an expression.

Example

Evaluate $3x^2 - y^2$ when $x = -2$ and $y = 4$

$3x^2 - y^2$	$= 3(-2)^2 - (4)^2$	Substitute.
	$= 3(4) - (16)$	Simplify the exponents.
	$= 12 - 16$	Multiply.
	$= -4$	Subtract.

The computer notation for an algebraic expression is different from traditional algebraic notation. A variable in the computer can involve two or more characters. In the computer, AX refers to a single variable. In algebra, AX symbolizes the product of the variables A and X.

Factors in computer notation must be separated by an asterisk, *. Algebraic notation for the product of 3, x, and y is $3xy$. Computer notation for the same product is $3 * x * y$.

Division in algebra is shown as a fraction or with a division symbol (÷). Division in computer notation is shown with a slash, /. Parentheses must be used if either the numerator or denominator involves more than one term.

Example

$\dfrac{3a - 2}{a + 4}$ is a rational expression in algebra. To write the same expression in computer notation, parentheses must be used in both the numerator and denominator: $(3 * a - 2)/(a + 4)$.

If the parentheses are omitted and the expression is written as $3 * a - 2/a + 4$, the hierarchy of operations dictates that $3 * a$ and $2/a$ be performed first. Then the additions and subtractions are performed. Hence, $(3 * a - 2)/(a + 4)$ does not have the same meaning as $3 * a - 2/a + 4$.

Addition and subtraction use the same symbols in algebraic and computer notation. Terms must be separated by either + or −.

Raising to a power is accomplished by the use of an arrow, ↑, in computer notation. Half-spaces are not available in computer printing without special instructions. The exponent affects only the factor directly preceding it unless parentheses are inserted. Thus a^2 is written as a↑2 in computer notation. Radicals are written by using the decimal equivalent of the fractional exponent representing the radical. The fractional quantity can be used if enclosed in parentheses.

Example

$$3\sqrt{a} = 3*a\uparrow.5 = 3*a\uparrow(1/2)$$

$$\sqrt{3a} = (3*a)\uparrow.5 = (3*a)\uparrow(1/2)$$

EXERCISES

In each of the following expressions, identify the terms, the factors of each of the terms, and the coefficient of x.

1. $3xy + 4a^2$
2. $4ax + 3y^2$

Given: $a = 2$, $b = -3$, $x = -4$, and $y = 5$. Evaluate each of the following and write the expression in computer notation.

3. $3b - 2a$
4. $3y - 2x$
5. $abx + y$
6. $x + aby$
7. $\dfrac{a - b}{y} + x$
8. $\dfrac{2x + 3}{y} + ab$
9. $3x^2 - by$
10. $3y^2 + ab$
11. $(a - b)^2 + xy$
12. $(x - y)^2 + ab$
13. $-b^2 + 3xy$
14. $-x^2 + 4ab$
15. $\dfrac{x - y}{a - b}$
16. $\dfrac{y - x}{b + a}$
17. $2a[b - x(3 + y)]$

18. $b[x - a(y - 3)]$
19. $3[(a + y)^2 - (by)]$
20. $-[(x + y)^3 - (ab)]$

Answers

2. $4ax$ and $3y^2$ are terms; $4a$ is the coefficient of x; 4, a, x, $4a$, $4x$, ax, and $4ax$ are factors of $4ax$; 3, y, $3y$, y^2, and $3y^2$ are factors of $3y^2$.
4. 23; 3*y - 2*x
6. -34; x + a*b*y
8. -7; (2*x + 3)/y + a*b
10. 69; 3*y↑2 + a*b
12. 75; (x - y)↑2 + a*b
14. -40; -x↑2 + 4*a*b
16. -9; (y - x)/(b + a)
18. +24; b*[x - a*(y - 3)]
20. -7; -[(x + y)↑3 - a*b]

6.2 SIMPLIFICATION

The computer cannot do algebra. For example, there is no command of *factor*. It is the job of the programmer to simplify expressions before entering them into programs. Simplification can result in substantial savings of time in a computer run, and time carries a significant money value.

Suppose the algebraic expression $-3x + 2[x - 3x(x^2 - 1)]$ is to be evaluated for numerous values of x. Simplify the expression, starting within the inside parentheses:

$$\begin{aligned} -3\{x + 2[x - 3x(x^2 - 1)]\} &= -3[x + 2(x - 3x^3 + 3x)] \\ &= -3(x + 2x - 6x^3 + 6x) \\ &= -3(9x - 6x^3) \\ &= -27x + 18x^3 \end{aligned}$$

The original expression had nine operations (square, subtract, multiply twice, subtract, multiply, add, multiply, and take the inverse). The simplified expression has five operations (cube, multiply twice, take the inverse, and add). For numerous values of x, substantial savings occur using the simplified expression.

The distributive property is involved in many simplification problems. For example, $3(x + y) = 3x + 3y$. The 3 is distributed to each term within the parentheses. Also, $3(xy + a) = 3xy + 3a$. Here, xy is a term and 3 is multiplied by the term, not by individual factors. The product $(3 + x)(a + 2)$

also utilizes the distributive property in its simplification: $(3 + x)(a + 2) = 3(a + 2) + x(a + 2) = 3a + 6 + ax + 2x$. In this case, the factored form is far more efficient than the expanded form as far as number of operations involved. The factored form involves two additions and one multiplication. The expanded form involves three multiplications and three additions.

There is no set rule for determining the most efficient form of an expression; the two forms must simply be compared.

Factoring is a process that converts an algebraic addition into a product. A factor is an element of a product. Factoring is the process of determining the elements of a product. As seen in the preceding example, the factored form $(3 + x)(a + 2)$ is a more efficient form than the expanded form. Factoring is also useful in solving equations and simplifying fractions.

The monomial factor, a single-term factor, occurs whenever every term of an expression contains a common multiple.

Example

$2a + 4$ has a common monomial factor of 2 because every term contains a factor of 2: $2a + 4 = 2(a + 2)$.

Example

$2ab + 3a$ has a common multiple of a and is factored as $a(2b + 3)$.

The difference of two squares can always be factored into the sum of the square roots of the terms times the difference of the square roots of the terms.

Example

$a^2 - b^2$ is the difference of two perfect squares and can be factored as $(a + b)(a - b)$.

Example

$9a^3 - 36a$ has a common factor of $9a$, so it equals $9a(a^2 - 4)$. The factor $a^2 - 4$ can be factored, and the entire expression becomes $9a(a + 2)(a - 2)$.

Trinomials can sometimes be factored as the product of two binomials. There is no fixed rule about how to determine the factors. Success depends on practice using trial and error.

Example

$$a^2 - 13ab - 30b^2 = (a - 15b)(a + 2b)$$

Example

$$a^2b - 19ab + 48b = b(a^2 - 19a + 48) = b(a - 16)(a - 3)$$

Occasionally common factors are binomials rather than monomials.

Example

$$9a(a - b) + 13(a - b) = (9a + 13)(a - b)$$

Example

$$3a - 9b + ab - 3b^2 = 3(a - 3b) + b(a - 3b) = (3 + b)(a - 3b).$$

Factoring is the reverse of the distributive property in mathematics; instead of distributing, the expression is being regrouped! Factoring, if all common factors are removed, produces a unique product.

EXERCISES

Perform the indicated operations and simplify by collecting similar terms.

1. $(x^2 - 2x + 3) - (2x^2 + 7x - 1)$
2. $(a^2 + 3x - 7) - (a^2 + 5x - 2)$
3. $(a - b) - (b - c) - (a - c)$
4. $(a + b + c) - (b - c) - (a + c)$
5. $2a(x + y)$
6. $3a(y - z)$
7. $a(2a - b)$
8. $b(2a - b)$
9. $(x + 5)(x - 3)$
10. $(x - 5)(x + 3)$
11. $(2a - 5)(3a + 1)$
12. $(2a + 5)(3a - 1)$
13. $-x\{x - 3[x(x^2 + 1)]\}$
14. $-a\{a^2 - 2[a(a - 1)]\}$

Write each of the following in factored form.

15. $5a^3 - 15a$
16. $3ab^2 + 11a^2b$
17. $a(7 - b) + b(7 - b)$
18. $b(x - 6) + a(x - 6)$
19. $a^2 - 16x^2$
20. $25b^2 - 49n^2$
21. $(x - 3)^2 - 25$
22. $(a - b)^2 - b^2$
23. $100a^2 - (b + c)^2$
24. $9c^2 - (a - b)^2$

25. $6a^2 - 19a - 7$
26. $3a^2 - 19a - 14$
27. $ax^2 - 9ax + 18a$
28. $5x^2 - 45x + 90$
29. $2a - ab + 4 - 2b$
30. $ax + ay - bx - by$

Answers

2. $-2x - 5$
4. c
6. $3ay - 3az$
8. $2ab - b^2$
10. $x^2 - 2x - 15$
12. $6a^2 + 13a - 5$
14. $a^3 - 2a^2$
16. $ab(3b + 11a)$
18. $(b + a)(x - 6)$
20. $(5b - 7n)(5b + 7n)$
22. $a(a - 2b)$
24. $(3c - a + b)(3c + a - b)$
26. $(3a + 2)(a - 7)$
28. $5(x - 3)(x - 6)$
30. $(a - b)(x + y)$

6.3 ALGEBRAIC FRACTIONS

There are three basic operations involved in fractions: simplification, addition and subtraction, and multiplication and division. Each of these operations has a set of procedures to guarantee correct results. The process used in arithmetic problems involving fractions can be used as a guide for algebraic fractions.

Simplification involves eliminating common factors in the numerator and denominator of a fraction

Procedure for Simplification

1. Factor both the numerator and denominator.
2. Reduce common factors to 1 (cancel like factors).
3. Write the remaining factors in their original locations. Expand if desired.

Example

Simplify $\frac{17}{51}$.

$$\frac{17}{51} = \frac{17}{17} \cdot \frac{1}{3} = (1)\frac{1}{3} = \frac{1}{3}$$

Example

Simplify $\frac{3a^2b^3}{27a^3b^2}$.

$$\frac{3a^2b^3}{27a^3b^2} = \frac{3a^2b^2b}{3a^2b^29a} = \frac{b}{9a}$$

Example

Simplify $\frac{2a - 5}{2a^2 - 3a - 5}$.

$$\frac{2a - 5}{2a^2 - 3a - 5} = \frac{(2a - 5)(1)}{(2a - 5)(a + 1)} = \frac{1}{a + 1}$$

The most common error in reducing fractions occurs when Step 1 is omitted. Only factors can be canceled.

Addition and subtraction of fractions involves finding a common denominator and writing each fraction with the common denominator.

Procedure for Addition and Subtraction

1. Determine a common denominator. A least common denominator makes the computations easier because the individual fractions are in lowest terms, but it is not essential for a correct answer.
2. Write each fraction with the common denominator determined in Step 1. This is accomplished by multiplying the numerator and denominator of each fraction by the factors missing in the original denominator.
3. Combine the numerators using the operations indicated. If a subtraction is indicated, each term of the numerator following the subtract sign is affected by the subtraction.
4. Write the combined simplified numerator over the common denominator determined in Step 1.
5. If the fraction can be simplified, reduce it to lowest terms.

Example

Add: $\frac{3}{a} + \frac{1}{b}$.

Step 1. The common denominator is $(a)(b)$.

Step 2. $\frac{3}{a} = \frac{3b}{ab}$; $\frac{1}{b} = \frac{a}{ab}$

Step 3. $\frac{3b}{ab} + \frac{a}{ab} = \frac{3b + a}{ab}$

Step 4. $\frac{3b + a}{ab}$ cannot be reduced.

Example

Subtract: $\frac{3}{a^2 - 4} - \frac{4}{a - 2}$.

Step 1. $a^2 - 4 = (a - 2)(a + 2)$ and $a - 2$ is prime; the common denominator is $(a + 2)(a - 2)$.

Step 2. $\frac{3}{a^2 - 4} = \frac{3}{(a - 2)(a + 2)}$; $\frac{4}{a - 2} = \frac{4(a + 2)}{(a - 2)(a + 2)}$

Step 3. $\frac{3}{(a - 2)(a + 2)} - \frac{4(a + 2)}{(a - 2)(a + 2)} = \frac{3 - 4(a + 2)}{(a - 2)(a + 2)}$

Step 4. $\frac{3 - 4a - 8}{(a - 2)(a + 2)} = \frac{-4a - 5}{(a - 2)(a + 2)}$, which cannot be reduced.

Multiplication and division are inverse operations; they utilize similar procedures. Division requires the additional step of inverting the divisor before the multiplication procedure is followed.

Procedure for Multiplication and Division

1. Invert any divisors.
2. Factor the expressions in each numerator and denominator.
3. Write as a single fraction. All factors of numerators become factors of the desired numerator and all factors of denominators become factors of the desired denominator.
4. Divide out like factors in the numerator and denominator.
5. Write the remaining factors in their appropriate location for the fractional answer. The expressions can be multiplied out or left in factored form. If all factors of the numerator or denominator cancel, a factor of 1 remains.

Example

Divide: $\frac{5a}{8b} \div \frac{10}{2b}$.

Step 1. Because this is a division problem, invert $10/2b$.

$$\frac{5a}{8b} \div \frac{10}{2b} = \frac{5a}{8b} \cdot \frac{2b}{10}$$

Steps 2-5. Factor both the numerators and denominators.

$$\frac{5a}{8b} \cdot \frac{2b}{10} = \frac{5 \cdot 2 \cdot a \cdot b}{2 \cdot 2 \cdot 2 \cdot 5 \cdot 2 \cdot b} = \frac{a}{8}$$

Example

$$\frac{a^2 - a - 6}{a^2 - 16} \div \frac{a - 3}{a + 4} = \frac{a^2 - a - 6}{a^2 - 16} \cdot \frac{a + 4}{a - 3}$$

$$= \frac{(a - 3)(a + 2)(a + 4)}{(a - 4)(a + 4)(a - 3)}$$

$$= \frac{a + 2}{a - 4}$$

These are the basic concepts for fractions. Additional algebraic techniques are sometimes needed to perform some problems because of special features of the algebraic expressions.

Example

$$\frac{a^2 - 7a + 12}{2a + 3} \cdot \frac{9 - 4a^2}{3 - a} = \frac{(a - 3)(a - 4)}{(2a + 3)} \cdot \frac{(3 - 2a)(3 + 2a)}{(3 - a)}$$

It is necessary to recognize that $2a + 3 = 3 + 2a$ (by the commutative property) and reduce. The factors $a - 3$ and $3 - a$ are not the same, but they are related: One is the additive inverse of the other; so $a - 3 = -(3 - a)$. Such factors in a fraction reduce to -1. Therefore, $(a - 3)$ and $(3 - a)$ reduce to the factor -1. The factor of -1 can be written in either the numerator or denominator but not in both.

$$\frac{(a - 4)(3 - 2a)}{-1} = (-1)(a - 4)(3 - 2a) = 2a^2 - 11a + 12$$

Division by 0 is undefined, and because fractions represent a division operation, care must be taken not to have a value of 0 in the denominator. It is wise to determine the values of the variable that will produce a 0 in the denominator and exclude those values from the problem.

Example

$\dfrac{x - 2}{x^2 - 2x}$ is undefined when $x^2 - 2x = 0$, which occurs when $x = 0$ or $x = 2$. Therefore the problem has the restriction that $x \neq 2$ or $x \neq 0$.

For all other values of x, the original fraction reduces to $\dfrac{1}{x}$.

EXERCISES

Reduce each fraction to lowest terms. Give any restrictions on the variables.

1. $\frac{5x^2}{2x}$
2. $\frac{6x^3}{9x^2}$
3. $\frac{24x^2y^3}{-8xy^2}$
4. $\frac{-32x^3y^2}{4xy}$
5. $\frac{3a^3 - 15a^2b}{6a}$
6. $\frac{2x^2y - 8xy^2}{4xy}$
7. $\frac{x - y}{y - x}$
8. $\frac{b - a}{a - b}$
9. $\frac{4x^2 - 4y^2}{4x + 4y}$
10. $\frac{3a - 3b}{3a^2 - 3b^2}$

Add or subtract. Give any restrictions on the variables.

11. $\frac{b}{a} + \frac{2}{b}$
12. $\frac{1}{f} + \frac{1}{n}$
13. $4 - \frac{1}{1 - a}$
14. $\frac{6}{1 - a} - 2$
15. $\frac{4}{a^2b} - \frac{7}{ab}$
16. $\frac{9}{ab} - \frac{4a}{b^2}$
17. $\frac{2}{x - a} - \frac{3}{x + b}$
18. $\frac{3}{x + a} - \frac{2}{x - b}$
19. $\frac{b}{ab - a^2} - \frac{a}{b^2 - ab}$
20. $\frac{b}{ab - ac} + \frac{a}{b^2 - bc}$
21. $\frac{a}{a - b} + \frac{1}{b^2 - a^2} + \frac{b}{a + b}$
22. $\frac{3}{x - 2} + \frac{x}{4 - x^2} + \frac{1}{x + 2}$

Multiply or divide. Give any restrictions for the variables in each problem.

23. $\frac{3a^2b^2c^3}{5} \div \frac{9a^3b^2c}{10}$
24. $\frac{2ab^2c^3}{7} \div \frac{8a^3b^2c}{21}$
25. $\frac{4x - 4y}{6t} \cdot \frac{18t^3}{x^2 - y^2}$
26. $\frac{3a^2 - 3b^2}{6x^2} \cdot \frac{10x}{a + b}$
27. $(x + 3) \cdot \frac{x}{x^2 + 4x + 3}$
28. $(1 - x) \cdot \frac{7x}{x^2 + 4x - 5}$
29. $\frac{x}{x - 2} \div \frac{2}{2 - x}$
30. $\frac{a}{b - a} \div \frac{b}{a - b}$
31. $\frac{x + 1}{x} \cdot \frac{2x}{x^2 + 6x + 5} \div \frac{1}{x + 5}$
32. $\frac{a - 3}{a} \div \frac{a^2 - 5a + 6}{a^2} \cdot \frac{2}{a - 2}$
33. For Exercises 17-31 odd, compare the number of operations involved in the original problem and the solution.

34. For Exercises 18-32 even, compare the number of operations involved in the original problem and the solution.

Answers

2. $\frac{2x}{3}, x \neq 0$

4. $-8x^2y, x \neq 0, y \neq 0$

6. $\frac{x-4y}{2}, x \neq 0, y \neq 0$

8. $-1, a \neq b$

10. $\frac{1}{a+b}, a \neq b, a \neq -b$

12. $\frac{n+f}{fn}, f \neq 0, n \neq 0$

14. $\frac{4+2a}{1-a}, a \neq 1$

16. $\frac{9b-4a^2}{ab^2}, a \neq 0, b \neq 0$

18. $\frac{x-3b-2a}{(x+a)(x-b)}, x \neq -a, x \neq b$

20. $\frac{b^2+a^2}{ab(b-c)}, a \neq 0, b \neq 0, b \neq c$

22. $\frac{3x+4}{(x-2)(x+2)}, x \neq 2, x \neq -2$

24. $\frac{3c^2}{4a^2}, a \neq 0, b \neq 0, c \neq 0$

26. $\frac{5(a-b)}{x}, x \neq 0, a \neq -b$

28. $\frac{-7x}{x+5}, x \neq 1, x \neq -5$

30. $\frac{-a}{b}, b \neq 0, b \neq a$

32. $\frac{2a}{(a-2)^2}, a \neq 0, a \neq 2, a \neq 3$

34.

Problem	18	20	22	24	26	28	30	32
Original	5	9	9	15	12	8	5	12
Answer	8	7	6	3	5	4	2	4

6.4 EXPONENTS AND RADICALS

Exponent. An *exponent* is the mathematical notation used to denote successive multiplications by the same factor. For example, $3^4 = 3 \cdot 3 \cdot 3 \cdot 3$. The computer also determines the value of an expression involving integer exponents by successive multiplications: 3^4 = 3↑4 = 3*3*3*3.

Radical. A *radical* is the symbol used in algebra to denote a root of a number. For example, $\sqrt{2} = 2^{1/2}$ is the square root of 2. The computer does not have a function or symbol for a radical in general, so the algebraic notation must be written with a fractional exponent. The exception to this is square root.

$\sqrt[4]{2^3} = 2^{3/4}$ = 2↑(.75) or 2↑(3/4) $\quad \sqrt[3]{\frac{1}{2}} = \sqrt[3]{2^{-1}} = 2^{-1/3}$ = 2↑(-1/3)

Expressions containing exponents can often be simplified before the expression is evaluated.

Rules for Simplifying Exponents

1. $x^a \cdot x^b = x^{a+b}$
2. $\dfrac{x^a}{x^b} = x^{a-b} \qquad (x \neq 0)$
3. $(x^a)^b = x^{ab}$
4. $\dfrac{x^a}{x^a} = 1 = x^0 \qquad (x^a \neq 0)$
5. $\dfrac{1}{x^a} = x^{-a} \qquad (x \neq 0)$
6. $\sqrt[a]{x^b} = x^{b/a} \qquad (x > 0)$
7. 0^0 is undefined.

Rules 1 and 2 are useful in simplification for computer expressions. Rule 3 is probably a costly rule in computer work. For example, $(x^3)^4$ involves 3 multiplications within the parentheses and then 4 multiplications for the exponent 4, making a total of 7 operations. If the expression x^{12} is substituted for the $(x^3)^4$ and the evaluation of the expression is desired for some value of x, 11 multiplications are involved. Rule 4 is very useful in simplification. Rule 5 makes it possible to write an expression without a denominator. Rule 6 is essential for computer work as a means of expressing radicals. Rule 7 is understood by the computer and will give an illegal quantity message.

Another place where exponents are found is in calculator work whenever the display or printout exceeds nine digits. Scientific notation is then used to express the number: $123{,}000{,}000{,}000{,}000 = 1.23 \times 10^{14}$ in scientific notation, which is printed as 1.23E+14 in the display of the calculator.

General Rules for Simplifying Expressions Involving Exponents and Radicals for Computer Notation

1. Write each factor only once in the problem.

Example

$$x^3 \cdot x^4 = x^7 \qquad x\sqrt{x} = x^{3/2}$$

2. Reduce numerical factors as much as possible.
3. Replace all radical signs with fractional exponents.

The use of denominators or negative exponents is a matter of individual preference. Exercises will give practice in both notations.

Recall that exponents affect only the factor directly preceding the exponent unless there are parentheses to indicate otherwise.

Example

$$-2^4 = -(2 \cdot 2 \cdot 2 \cdot 2) = -16$$

EXERCISES

Use scientific notation to compute each of the following. Write the answer in the notation that would be used in computer display.

1. $\dfrac{(.00004)^2 \times 500{,}000{,}000{,}000}{.0020}$
2. $\dfrac{(.000036) \times (20{,}000{,}000)^2}{1200}$
3. $\dfrac{(7000000)(120000000)}{.004}$
4. $\dfrac{.000084}{.00007 \times .03}$

Simplify each of the following and write the simplified expression in computer notation. Use only positive exponents.

5. $\dfrac{2x^7}{6x^{-3}}$
6. $\dfrac{15x^{-3}}{5x^7}$

7. $(36x^6)\left(\frac{x^3}{12}\right)$
8. $\left(\frac{54}{x^4}\right)\left(\frac{x^{-2}}{9}\right)$
9. a^4x^4
10. b^3x^6
11. $\sqrt[3]{x^2y^3}$
12. $\sqrt[5]{x^{10}y^6}$
13. $\frac{1}{x^{-2}}$
14. $\frac{x^{-2}}{y^{-3}}$
15. $10^n \cdot 10$
16. $2^x \cdot 2^3$
17. $(a^2b^3)^n$
18. $(x^{-2}y^3)^{n+1}$
19. $\frac{18x^3y^8z^5}{9x^5y^5z^5}$
20. $\frac{-2^2x^4y^6}{12x^2y^4}$
21. $2^6 \cdot 4^3$
22. $4^4 \cdot 16^2$
23. $9x^4 \cdot 3^2$
24. $\sqrt{x} \cdot \sqrt[4]{x^2}$
25. $\sqrt{3} \cdot \sqrt[4]{3^2}$

Answers

2. 12,000,000
4. 40
6. 3/x↑10
8. 6/x↑6
10. (b*x↑2)↑3
12. x↑2*y↑(6/5)
14. y↑3/x↑2
16. 2↑(x+3)
18. (y↑3/x↑2)↑(n + 1)
20. x↑2*y↑2/(-3)
22. $(16)^4$, or $(4)^8$
24. x

6.5 LINEAR EQUATIONS

A *linear equation* is an algebraic sentence containing an equality sign and a variable of degree 1. For example, $2x - 1 = 8a^2 + 5$ is a linear equation in x and $x^2 + 3x = 5x - 7 + x^2$ is a linear equation because the x^2-terms drop out when the equation is simplified. From these examples it is obvious the variable must be identified to determine the degree of an equation and simplification must be completed before the degree is determined.

The computer cannot solve equations. It is possible to substitute successive values of a variable into a statement to check the validity of a statement. This is not a practical way to solve an equation. The algebraic manipulations must be performed by the programmer; the computer performs arithmetic operations.

Linear equations have one solution unless the equation involves a contradiction, in which case there are no solutions, or a tautology, where there are an infinite number of solutions.

Steps for Solving a Linear Equation

1. Clear the statement of parentheses.
2. Clear the statement of denominators by multiplying each term by the LCD (least common denominator) of the equation.
3. Put all terms containing the unknown on one side of the equation and all other terms on the other side.
4. If there is more than one term containing the unknown, combine the terms using the distributive property.
5. Divide both sides of the equation by the coefficient of the unknown.

A linear equation to be programmed on a computer can involve more than one variable. The problem is solved for one of the variables in terms of the others present. Specific values are then substituted for the letters, and corresponding values are determined for the unknown variable.

Example

A weather station requires a printout of the Celsius readings corresponding to Fahrenheit readings. The station wishes conversions for Fahrenheit readings of 10°F to 100°F. The formula is $F = \frac{9}{5}C + 32$.

If Celsius readings are required, the formula must be solved for C.

There are no parentheses; however it is necessary to clear the denominator 5.

$5F = 9C + 160$ Multiply both sides by 5.

$5F - 160 = 9C$ Grouping C on one side and everything else on the other.

$\frac{5F - 160}{9} = C$ Now divide both sides by the coefficient of C.

Written in computer notation, this is (5*F − 160)/9. A flowchart for the computer printout of this problem is as follows.

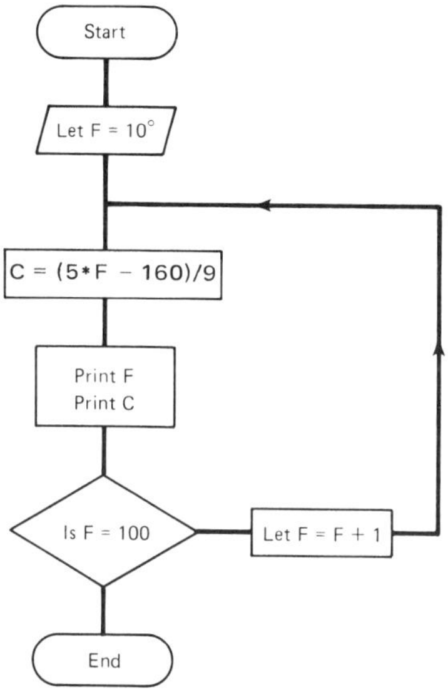

Computer work demands a knowledge of solving equations. The exercises will give some numerical equations to use for practicing the procedure. Some literal equations are also given. Each equation will indicate the variable for which to solve.

Example

$4a - 5y = 3(y - 2a)$

Step 1. $4a - 5y = 3y - 6a$

Step 2. $4a + 6a = 3y + 5y$

Step 3. $10a = 8y$

Step 4. $\frac{10a}{8} = y$

$\frac{5a}{4} = y$

EXERCISES

Solve each equation for x using the procedure outlined in this section. If there are any restrictions on the variable, indicate the restrictions.

1. $3x - 15 = 17 - 5x$

2. $8x - 5 = 20 + 3x$
3. $2(x - 3) = 7x + 9$
4. $3x + 5 = 6(x + 1)$
5. $3 - 2(x - 4) = 6x - 1$
6. $8 - x(3 - x) = x(6 + x)$
7. $\frac{7x}{6} = 21$
8. $\frac{2x}{3} = 28$
9. $\frac{3x}{10} - \frac{2x}{5} = \frac{7}{3}$
10. $\frac{3x}{7} - \frac{1}{3} = \frac{5x}{6}$
11. $\frac{2x}{x^2 - 4} - \frac{3}{x - 2} = \frac{4}{x + 2}$
12. $\frac{x}{x^2 - 1} - \frac{1}{x + 1} = \frac{5}{x - 1}$
13. $\frac{7}{2x - 5} - \frac{2}{3} = \frac{5}{2}$
14. $\frac{2}{7} - \frac{1}{x + 5} = \frac{1}{14}$

Solve each literal equation for the variable indicated. Write the solution in computer notation.

15. $A = \frac{1}{2}h(b + B)$; solve for h.
16. $A = P(1 + rt)$; solve for r.
17. $L = a + (n - 1)\,d$; solve for n.
18. $S = \frac{P}{1 + dt}$; solve for d.
19. $S = \frac{a - rd}{1 - r}$; solve for r.
20. $\frac{1}{R} = \frac{1}{R_1} + \frac{1}{R_2}$; R, R_1, and R_2 denote different variables, solve for R_1.
21. $\frac{1}{f} = \frac{1}{p} + \frac{1}{x}$; solve for f.
22. $A = \frac{1}{2}h(b_1 + b_2)$; solve for b_1.
23. $V = \frac{1}{3}h\left(\frac{r_1 + r_2}{2}\right)$; solve for r_1.
24. $a = \frac{c - b}{c - k}$; solve for c.

25. $b - a = \dfrac{d - b}{d + a}$; solve for b.

Answers

2. $x = 5$
4. $x = \dfrac{-1}{3}$
6. $x = \frac{8}{9}$
8. $x = 42$
10. $x = -\frac{14}{17}$
12. $x = -\frac{4}{5}, x \neq 1, x \neq -1$
14. $x = -\frac{1}{3}, x \neq -5$
16. $r = \dfrac{A-P}{Pt}$ = (A-P)/(P*t)
18. $d = \dfrac{P-s}{st}$ = (P-s)/(s*t)
20. $R_1 = \dfrac{RR_2}{R_2-R}$ = R*R$_2$/(R$_2$-R), $R, R_1, R_2 \neq 0$
22. $b_1 = \dfrac{2A-hb_2}{h}$ = (2*A-h*b$_2$)/h
24. $c = \dfrac{ak-b}{a-1}$ = (a*k-b)/(a-1), $c \neq k$

6.6 LINEAR INEQUALITIES

In many applications of mathematics there is a need to work with algebraic inequalities. There are five possible inequalities:

Symbol	Meaning	Computer notation
$>$	Greater than (not equal to or less than)	>
$<$	Less than (not equal to or greater than)	<
$\geq$	Greater than or equal to (not less than)	>=
$\leq$	Less than or equal to (not greater than)	<=
$\neq$	Not equal to (greater than or less than)	<>

When a linear algebraic statement involves one of these symbols, it is called a linear inequality. Most of the algebraic manipulations used in solving linear equations work with inequalities. The exception is the operation of multiplication or division by a negative number. In a statement involving

$>$ or $<$, the direction of the inequality must be reversed when the algebraic manipulation is performed.

Example

Solve $\frac{x-2}{2} < x + \frac{3x-1}{2}$ for x.

$$\frac{x-2}{2} < x + \frac{3x-1}{2}$$

$x - 2 < 2x + 3x - 1$ Clearing the denominator.

$x - 2 < 5x - 1$ Combine like terms.

$x - 5x < -1 + 2$ Grouping the terms.

$-4x < 1$ Simplify.

$x > -\frac{1}{4}$ Divide by -4. Reverse the direction of the inequality.

Occasionally two inequalities will be given simultaneously, where a solution is required for both. If *and* connects the two inequalities, the final solution is the intersection of the two separate solutions. If *or* connects the inequalities, the final solution is the union of the two separate solutions.

Example

Solve $4x - 5 > 3x - 2$ and $3x - 11 < x + 9$ for x.

$4x - 5 > 3x - 2$ and $3x - 11 < x + 9$

$x > 3$ $\quad$ $2x < 20$

$x < 10$

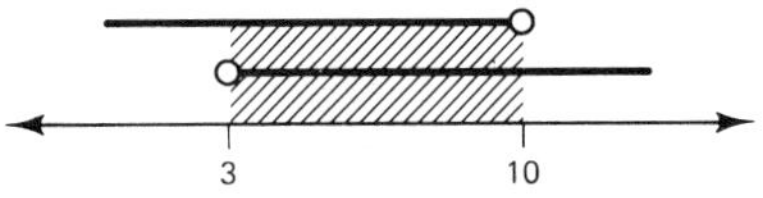

Answer: $3 < x < 10$

Example

Solve $x - 1 > 2$ or $x - 1 < -2$ for x.

$x - 1 > 2$ or $x - 1 < -2$

$x > 3$ $\quad$ $x < -1$

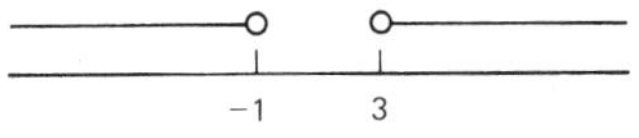

Answer: All real numbers except -1 to 3, inclusive.

Great care must be taken in solving linear inequalities that involve several letters.

Example

$$ax > 10$$

$$x \; ? \; \frac{10}{a}$$

If $a > 0$, $x > \frac{10}{a}$ If $a > 0$, the inequality sign remains the same.

If $a < 0$, $x < \frac{10}{a}$ If $a < 0$, the inequality sign is reversed.

If $a = 0$, the statement is undefined.

In a flowchart for this solution, the sign of a must be checked.

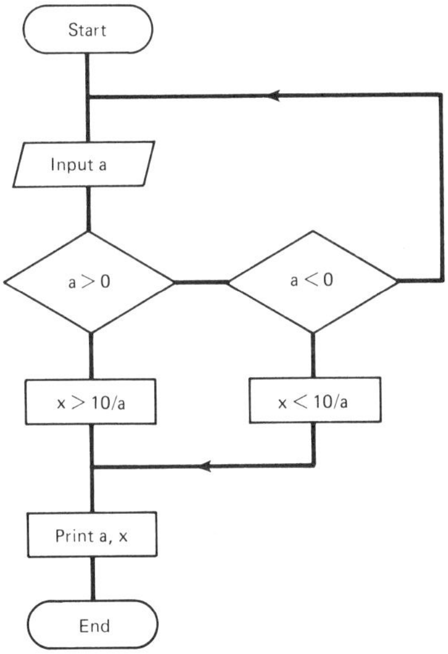

EXERCISES

Solve each inequality for x.

1. $3x - 2 > 4x + 7$
2. $4x - 1 < 6x + 5$
3. $\frac{x}{3} - 4 \geq 2x - 1$

4. $x - 5 < \frac{1 - 3x}{4}$

5. $\frac{3x + 1}{5} \leq \frac{3x - 1}{4}$

6. $\frac{4 - 3x}{2} > \frac{x + 1}{3}$

Solve each of the following pairs of inequalities. Sketch the solution on a number line.

7. $\frac{3x + 2}{2} < \frac{x + 5}{5}$ and $\frac{5x - 2}{4} > \frac{x - 2}{3}$

8. $3x + 2 > x + 5$ or $\frac{x - 2}{2} < \frac{3x - 1}{3} + x$

Solve each of the following literal inequalities for x and state any restrictions.

9. $ax + by \geq 120$

10. $\frac{x}{a} + \frac{y}{b} < 100$

11. $a(y - b) \geq b(x + a)$

12. $\frac{x - y}{a} < b$

Answers

2. $x > -3$
4. $x < 3$
6. $x < \frac{10}{11}$
8. $x > -\frac{4}{9}$
10. $x < \frac{100ab - ay}{b}$ if $a > 0$; $x > \frac{100ab - ay}{b}$ if $a < 0$; x is undefined if $a = 0$.
12. $x < ab + y$ if $a > 0$, $x > ab + y$ if $a < 0$; x is undefined if $a = 0$.

6.7 QUADRATIC EQUATIONS

A quadratic equation is an algebraic statement containing an equality sign and a variable of degree 2. The general form of a quadratic equation is $ax^2 + bx + c = 0$. The only term necessary for the equation to be quadratic is ax^2.

Example

$$2x^2 + 7 = 0 \qquad a = 2, b = 0, c = 7$$

$$x^2 - 2x = 0 \qquad a = 1, b = -2, c = 0$$

$$x^2 - 2x + 7 = 0 \qquad a = 1, b = -2, c = 7$$

$$7x^2 = 0 \qquad a = 7, b = 0, c = 0$$

Notice that all quadratics are set equal to 0 before a, b, and c are identified. If the quadratic $2x^2 + 3x = 4$ is given, it is necessary to write it as $2x^2 + 3x - 4 = 0$ before identifying $a = 2$, $b = 3$, and $c = -4$.

The most familiar method of solving a quadratic equation is factoring. This is the quickest and easiest way if it works.

Example

Solve $x^2 - x = 6$.

Before factoring, one side of the equation must be 0.

$$x^2 - x - 6 = 0$$

$$(x - 3)(x + 2) = 0 \quad \text{Rewrite } x^2 - x - 6 \text{ as a product.}$$

The key to the factoring method is to write the equation as a product equal to 0. The only way two quantities can have a product of 0 is if at least one of the factors is 0. Therefore, either $x - 3 = 0$, or $x + 2 = 0$, or $x - 3$ and $x + 2$ are both equal to 0.

If $x - 3 = 0$, $x = 3$.
If $x + 2 = 0$, $x = -2$.
If $x - 3$ and $x + 2 = 0$, $x = 3$ and -2, which is impossible.

Therefore the solution to $x^2 - x = 6$ is $x = 3$ or $x = -2$. Generally quadratics have two solutions.

General Procedure for Solving Quadratic Equations by Factoring

1. Arrange all terms on one side of an equation with 0 on other side.
2. Write the terms in factored form.
3. Set each factor equal to 0 and solve for the value of the variable in that factor.

There are two deterrents to factoring:

1. Not all quadratics factor.
2. The computer does not respond to a command of factor.

There are three basic methods of solving a quadratic equation.

1. Factoring: This method is for quadratics that will factor; it is quick and easy; it is not programmable for computer work.
2. Completing the square: This method is good for all quadratics; it is lengthy and cumbersome; it is programmable.
3. Quadratic formula: This method is good for all quadratics; it is subject to arithmetic errors; it is programmable.

The factoring technique has been discussed. The only reason for suggesting the method of completing the square is to lead to the quadratic formula.

Quadratic equations of the type $x^2 = 16$ are often solved by taking the square root of both sides of the equation and getting $x = \pm 4$. This concept led mathematicians to develop the technique of completing the square. The equation $x^2 + 2x - 5 = 0$ cannot be factored. However, if this expression could be written $(\ )^2 =$ some number, the technique of taking the square root could be applied.

General Procedure for Solving Quadratic Equations by Completing the Square

1. Divide each term by the coefficient of x^2.
2. Move the numerical term to the other side of the equality.
3. Add the numerical term that produces a perfect square trinomial to each side of the equality.
4. Write the trinomial in factored form; simplify the numerical terms.
5. Take the square root of each side of the equality.
6. Solve for x.

Example

Solve $x^2 + 2x - 5 = 0$

Step 1. $\frac{x^2}{1} + \frac{2x}{1} - \frac{5}{1} = \frac{0}{1}$

Step 2. $x^2 + 2x = 5$

Step 3. $x^2 + 2x + \underline{1} = 5 + \underline{1}$

Step 4. $(x + 1)^2 = 6$

Step 5. $x + 1 = \pm\sqrt{6}$

Step 6. $x = \pm\sqrt{6} - 1$

In general practice this method can produce negative numbers under the radical. Square roots of negative numbers require extending the number system to complex numbers, which is not within the scope of this book.

The technique of completing the square is applied to the general quadratic equation

$$ax^2 + bx + c = 0$$

The steps correspond to the procedure outlined above.

Step 1. $\dfrac{ax^2}{a} + \dfrac{bx}{a} + \dfrac{c}{a} = \dfrac{0}{a}$

$$x^2 + \frac{b}{a}x + \frac{c}{a} = 0$$

Step 2. $x^2 + \dfrac{b}{a}x = -\dfrac{c}{a}$

Step 3. $x^2 + \dfrac{b}{a}x + \dfrac{b^2}{4a^2} = -\dfrac{c}{a} + \dfrac{b^2}{4a^2}$ $\dfrac{1}{2}$ of $\dfrac{b}{a}$ is $\dfrac{b}{2a}$; $\left(\dfrac{b}{2a}\right)^2 = \dfrac{b^2}{4a^2}$

Step 4. $x^2 + \dfrac{b}{a}x + \dfrac{b^2}{4a^2} = \left(x + \dfrac{b}{2a}\right)^2$

$$\frac{-c}{a} + \frac{b^2}{4a^2} = \frac{-4ac}{4a^2} + \frac{b^2}{4a^2} = \frac{-4ac + b^2}{4a^2}$$

$$\left(x + \frac{b}{2a}\right)^2 = \frac{-4ac + b^2}{4a^2}$$

Step 5. $x + \dfrac{b}{2a} = \pm\sqrt{\dfrac{-4ac + b^2}{4a^2}} = \dfrac{\pm\sqrt{-4ac + b^2}}{2a}$

Step 6. $x = \pm\dfrac{\sqrt{-4ac + b^2}}{2a} - \dfrac{b}{2a} = \dfrac{-b \pm\sqrt{b^2 - 4ac}}{2a}$ Quadratic formula

The real use of the method of completing the square is in generating the general formula that will solve all quadratic equations.

Given a quadratic equation in the form $ax^2 + bx + c = 0$ where a, b, and c represent any real numbers and $a \neq 0$:

$$x = \frac{-b \pm \sqrt{b^2 - 4ac}}{2a}$$

Example

Solve $2x^2 - x = 7$.

For any method of solution, the equation must be set equal to 0, which gives $2x^2 - x - 7 = 0$. This does not factor. By the quadratic formula, with $a = 2$, $b = -1$, and $c = -7$,

$$x = \frac{1 \pm \sqrt{1 - (4)(2)(-7)}}{2(2)} = \frac{1 \pm \sqrt{57}}{4} \qquad x_1 = 2.14 \qquad x_2 = -1.64$$

These are approximate answers. Any desired accuracy can be expressed.

The expression under the radical sign, $b^2 - 4ac$, is called the *discriminant*. The value of this expression determines the nature of the roots.

If $b^2 - 4ac < 0$, the roots are complex numbers.
If $b^2 - 4ac = 0$, the roots are equal; this is called a *double root.*
If $b^2 - 4ac > 0$, the roots are rational if $b^2 - 4ac$ is a perfect square; the roots are irrational if $b^2 - 4ac$ is not a perfect square.

In a flowchart for the solutions of a quadratic equation, it is wise to look at the discriminant first to determine the nature of the roots. In many applications, especially business, complex roots are not meaningful.

If complex roots were needed, the decision box involving $b^2 - 4ac$ would lead to instructions to convert the negative discriminant into a form the computer could handle, the formula would be applied, and appropriate results would be recorded.

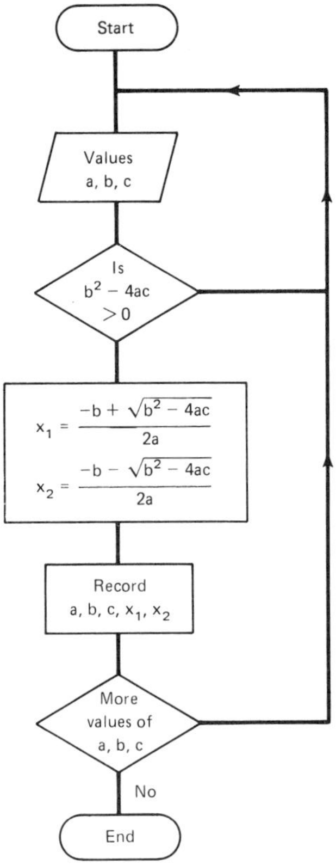

EXERCISES

In each of the following quadratic equations, put the equation in the form $ax^2 + bx + c = 0$. Determine the value of the discriminant. If the equation has rational roots, factor and give the solutions. If the equation has irrational roots, use the quadratic formula to determine the solution.

1. $2x^2 = 3x$
2. $9x^2 = 12x$
3. $2x^2 - 5x = 1$
4. $3x^2 - 5x = 1$
5. $9x^2 = 64$
6. $36x^2 = 25$
7. $16x^2 + 40x + 25 = 0$
8. $49x^2 - 28x + 4 = 0$
9. $2x^2 + 7 = 5 + 5x$
10. $3x^2 + 11 = 9 + 6x$
11. $3x^2 = 2x + 2$
12. $x^2 = 3x + 9$
13. $x^2 + 6x + 3 = 0$
14. $2x^2 + 6x + 3 = 0$
15. $6 - [(x - 3)(x - 1)] = 4$
16. $10 - [(x - 2)(x + 2)] = 6x$
17. $1.2x^2 - 4.3x - 5.5 = 0$
18. $3.4x^2 + 7.2x - 4.1 = 0$
19. $-x^2 + 8x = 1$
20. $-x^2 - x = -1$

Answers

2. Discriminant = 144; $x_1 = 0, x_2 = \frac{4}{3}$
4. Discriminant = 37; $x_1 = 1.85, x_2 = -.18$
6. Discriminant = 3600; $x_1 = \frac{5}{6}, x_2 = -\frac{5}{6}$
8. Discriminant = 0; $x_1 = x_2 = \frac{2}{7}$
10. Discriminant = 12; $x_1 = 1.58, x_2 = .42$
12. Discriminant = 45; $x_1 = 4.85, x_2 = -1.85$
14. Discriminant = 12; $x_1 = -2.37, x_2 = -.63$
16. Discriminant = 92; $x_1 = 1.80, x_2 = -7.80$
18. Discriminant = 107.6; $x_1 = .47, x_2 = -2.58$
20. Discriminant = 5; $x_1 = .62, x_2 = -1.62$

6.8 HIGHER-DEGREE EQUATIONS

The last two sections addressed the solution of linear and quadratic equations. The quadratic formula is ideal for computer use and solves any quadratic equation. However, if the degree of the equation is higher than two, no such formula exists. It is necessary to employ other techniques. Higher mathematics and calculus provide additional means of solving such equations. Without the benefit of more advanced mathematics, there are two methods available to approximate roots of higher-degree equations.

Bolzano Method

The Bolzano method is an iterative process that methodically approaches a root of an equation. Polynomial equations have graphs that are continuous, well-behaved curves. If two values of opposite sign are found for a polynomial equation, a root exists between the two x-values that produced these values of opposite signs. The Bolzano method provides a way to determine a root of an equation to the desired accuracy.

Example

Given the polynomial equation $x^3 + 2x^2 - x - 10 = y$. The equation formed by replacing y by 0 is $x^3 + 2x^2 - x - 10 = 0$. This is the equation to be solved, but the Bolzano method refers back to the equation involving x and y. By substituting x values in the original equation, y values can be found to correspond to the given x values.

When $x = 0, y = -10$
When $x = 1, y = -8$
When $x = 2, y = 4$

The sign change in the y value indicates there is a value of x between 1 and 2 that makes $x^3 + 2x^2 - x - 10 = 0$. Determine the midpoint between 1 and 2 as the first approximation x_1. To find the midpoint, add 1 and 2 together and divide by 2. $x_1 = (1 + 2)/2 = 1.5$; the y value if $x = 1.5$ is $(1.5)^3 + 2(1.5)^2 - (1.5) - 10 = -3.625$.

The accuracy desired in this problem is for the x value to produce a y value less than or equal to .01. x_2 is determined from 1.5 and 2 because of the sign change in y. $x_2 = (1.5 + 2)/2 = 1.75$. The corresponding y value is $-.265625$. The process continues until the desired accuracy is attained. A table is useful to record corresponding y values for each x approximation.

	x	y
a	1	-8
b	2	4
x_1	1.5	-3.625
x_2	1.75	-.2656
x_3	1.875	1.748
x_4	1.825	.915
x_5	1.7875	.31415
x_6	1.76875	.021696
x_7	1.75938	-.1226
x_8	1.7641	-.0506
x_9	1.7664	-.0145

$$x_2 = \frac{1.5 + 2}{2} = 1.75, \qquad \text{which gives a } y\text{-value of } -.265625$$

$x_3 = \dfrac{1.75 + 2}{2} = 1.875,$ which gives a y-value of 1.748

$x_4 = \dfrac{1.75 + 1.875}{2} = 1.825,$ which gives a y-value of .915

$x_5 = \dfrac{1.75 + 1.825}{2} = 1.7875,$ which gives a y-value of .31415

$x_6 = \dfrac{1.75 + 1.7875}{2} = 1.76875,$ which gives a y-value of .021696

$x_7 = \dfrac{1.75 + 1.76875}{2} = 1.7594,$ which gives a y-value of $-.1226$

$x_8 = \dfrac{1.7594 + 1.76875}{2} = 1.7641,$ which gives a y-value of $-.0506$

$x_9 = \dfrac{1.7641 + 1.76875}{2} = 1.7664,$ which gives a y-value of $-.0145$

At this point, a fair approximation for the root is 1.77. This decision could have been made back in the second step, when the small y-value was obtained for the x-value of 1.75. However this method does not allow for this type of reasoning; it is a process that continues until the desired accuracy has been obtained. The accuracy of this problem called for y to be less than or equal to .01 when y was expressed to hundredths.

The graphical picture of the process is shown in the figure.

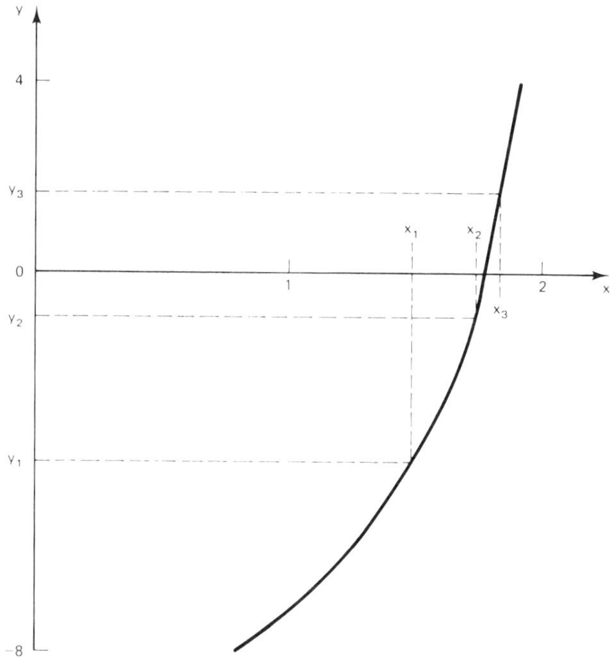

Flowchart for the Computer Use of the Bolzano Method

1. Pick preliminary values a and b where the y_a- and y_b-values have different signs.
2. $x_n = (a + b)/2$; calculate y_n using x_n in the original function.
3. Compare the signs of y_a and y_n.
4. $a = x_n$ or $b = x_n$; if the signs of y_a and y_n are alike, $b = x_n$. If the signs of y_a and y_n are unlike, $a = x_n$.
5. Calculate new x_n- and y_n-values; continue until the desired accuracy is achieved.

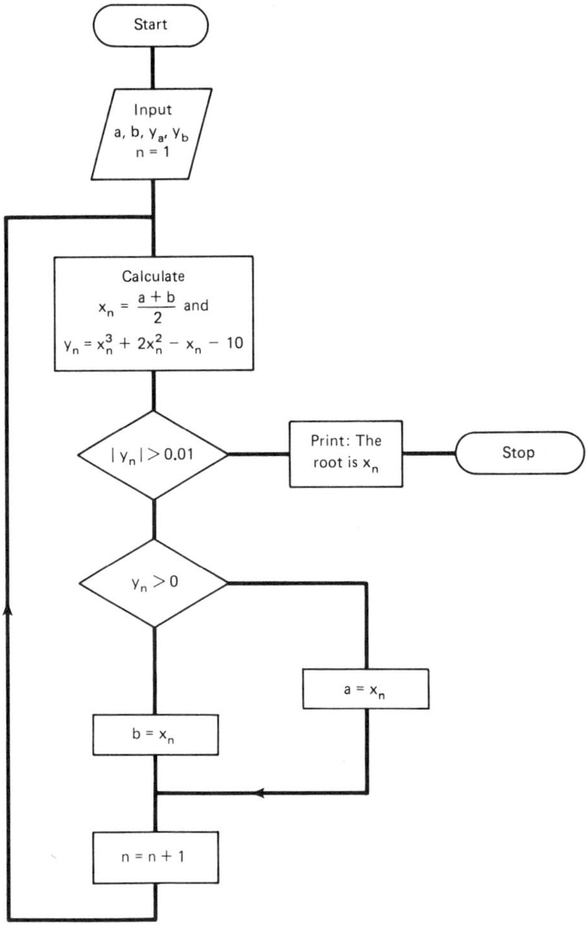

Root by False Position

The method of *root by false position* utilizes the concept of linear interpolation and the process gives an approximate root in fewer steps. The name comes from the graphical interpretation of the process.

Example

Consider the example just worked by the Bolzano method: $y = x^2 + 2x^2 - x - 10$ has a root between $x = 1$, where $y = -8$, and $x = 2$, where $y = 4$. By looking at the y-values, the root appears nearer 2 than 1 because 4 is closer to 0 than -8.

Arranging the values in tabular form helps set up the procedure.

		x	y		
1	$x_1 - 1$	1	-8	8	12
		x_1	0		
		2	4		

The middle line comes from naming the first approximate root x_1; the desired y value is 0. The numbers 8, 12, $x_1 - 1$, and 1 are the differences obtained in the subtractions indicated. Set up a proportion of these differences.

$$\frac{x_1 - 1}{1} = \frac{8}{12}$$

By finding cross products, this becomes $12x_1 - 12 = 8$, so $x_1 = 1.67$. Then 1.67 is substituted in the original equation: $(1.67)^3 + 2(1.67)^2 - (1.67) - 10 = -1.48$. We have made our first approximation for x, giving a y-value of -1.48. The x-value 1.67 replaces 1 and the y-value -1.48 replaces -8 in the above table.

		x	y		
.33	$x_2 - 1.67$	1.67	-1.48	1.48	5.48
		x_2	0		
		2	4		

$$\frac{x_2 - 1.67}{.33} = \frac{1.48}{5.48}$$

$$x_2 = 1.76$$

This process took two steps to do what the Bolzano method did in nine calculations. The flowchart is almost identical to the one used in the Bolzano method with the calculation $(a + b)/2$ replaced by

$$x_n = \frac{-y_a}{y_b - y_a}(b - a) + a$$

This method works for a polynomial of any degree. It is necessary to find values of x that give values of opposite sign for y.

EXERCISES

Solve each of the following polynomial equations for the root in interval indicated. Use the Bolzano and the false-position methods on each example. The accuracy desired is $|y| < .1$

1. $2x^3 + 3x^2 + 7x + 1 = y$; between -1 and 0
2. $2x^3 + x^2 - 7x + 2 = y$; between 1 and 2
3. $x^4 - x^3 - x^2 - x + 1 = y$; between 1 and 2
4. $3x^4 - 2x^3 + x^2 - x - 7 = y$; between 1 and 2
5. Use the polynomial function $y = x^3 - 20$ to find the cube root of 20.
6. Use the polynomial function $y = x^5 - 20$ to find the fifth root of 20.
7. Find the root of $y = x^2 - 4x + 2$ between 3 and 4. Compare to the solution by the quadratic formula.
8. Find the root of $y = x^2 - 4x + 2$ between 0 and 1. Compare to the solution by the quadratic formula.

Answers

2. 1.4375
4. 1.42
6. 1.82
8. .59

ADDITIONAL PROBLEMS

1. A gardener bought fertilizer for \$306. Had she bought 1 bag more for the same amount of money, the fertilizer would have cost her \$1 less per bag. How many bags did she buy?
2. A landscape architect bought some plants for \$60. Had he bought 4 fewer plants for the same money, he would have paid 50¢ more for each plant. How many plants did he buy?
3. An interior decorator had to pay income taxes of \$1920 plus 24% of the amount her taxable income exceeded \$20,000. Her tax bill was \$2280. What was her taxable income?
4. A professional athlete had to pay income taxes of \$40,000 plus 72% of the amount his taxable income exceeded \$82,000. His tax bill was \$43,600. What was his taxable income?
5. A water company changed its billing procedure from \$19.50 per 1000 gallons to \$1.75 per 100 gallons plus a service charge of \$25 per month. How much water must be purchased before the new monthly bill would be the same as the bill under the former rate?
6. A telephone company changed its billing procedure from \$19.50 per month to \$10.00 per month plus a service charge of 25¢ per local call. How many local calls would a household be allowed so the new monthly bill would not exceed the original billing rate?
7. A copying machine is installed in an office with a charge of \$50 per month plus 10¢ per copy for the first 5000 copies made, 8¢ per copy for the next 6000 copies and 5¢

per copy for all copies made in excess of 11,000 per month. Write the expression used to determine the amount due at the end of the month. Calculate the amount due if 14,867 copies are run.

8. A computer installation has the following rate structure for the equipment: base charge of \$1000 per month plus \$340 per hour for the first 10 hours of use, \$300 per hour for the next 25 hours, and \$200 for all hours in excess of 35. Write the expression used to determine the monthly charge. Calculate the charge for 30 hours of use.
9. An advertisement is printed on a sheet of paper 40 centimeters by 16 centimeters. The advertisement consists of an area of 385 square centimeters of printed material surrounded by a uniform margin. Find the width of the margin.
10. A circular is printed on a sheet of paper 9 inches by 11 inches. The advertisement consists of an area of 63 square inches of printed material surrounded by a uniform margin. Find the width of the margin.
11. A group of travelers chartered a bus for \$480 by each contributing equal amounts of money. If one more person joined the group, the contribution of each person would be \$1 less. How many people were in the original group?
12. A group of travelers chartered a tour for \$4800. Each person contributed an equal amount. If one more person joined the group, the contribution for each person would be \$20 less. What was the cost per person for the originally planned tour?
13. A container company is making boxes from rectangular sheets of cardboard by cutting out a square corner from each corner and bending up the sides to form a box. The cardboard has dimensions of 30 inches by 36 inches. The required box must have a volume of 2464 cubic inches. Find the dimensions of the box.
14. A container company is making boxes from rectangular sheets of cardboard by cutting out a square corner from each corner of the cardboard and folding up the sides to form a box. The required box must have a volume of 3432 cubic inches. The original dimensions of the rectangular sheet of cardboard was 38 inches by 34 inches. Find the dimensions of the box.
15. How much must a company earn to have \$150,000 left after paying taxes of 32% of earnings?
16. How much must a person earn to have an income of \$50,000 after taxes if the rate of taxation is 42%?

Answers

2. $x = 24$
4. \$87,000
6. 38 calls
8. x = no. of hours used.

$x \leq 10$	Charge = $1000 + 340x$
$10 < x \leq 35$	Charge = $1000 + 3400 + 300(x - 10)$
$35 < x$	Charge = $1000 + 3400 + 7500 + 200(x - 35)$
If $x = 30$,	Charge = $1000 + 3400 + 6000 =$ \$10,400

10. 1″

12. \$320
14. 26 inches $\times$ 22 inches $\times$ 6 inches
16. \$86,202

CHAPTER 6 REVIEW EXERCISES

Solve the following for x or the variable indicated.

1. $\frac{2}{x} + \frac{3}{2x} = 7$; state any restrictions on the variable x
2. $\frac{1}{2} + \frac{1}{x - 2} = 1$; state any restriction on the variable x.
3. $x - 3(2 - x) = 6$
4. $8 - 4[x - 3(2 - x)] = 0$
5. $\frac{x}{a} + 16b = \frac{x - a}{b}$; write the expression for x in computer notation.
6. $\frac{x - b}{a} - \frac{a + x}{b} = 7$; write the expression for x in computer notation.
7. $\frac{3}{f} + \frac{4}{p} = \frac{2}{q}$; solve for p.
8. $\frac{3}{R_1} + \frac{2}{R_2} = \frac{1}{R}$; solve for R.
9. $14x^2 - 3 = x$
10. $24x^2 - 13x = 7$
11. $7x^2 - x = 1$
12. $5x^2 - 2x = 6$
13. $1.3x^2 - 2.6x - 11.2 = 0$
14. $3.1x^2 - 6.2x = 21.1$
15. $3x - 2 > 6(1 + 2x)$
16. $\frac{5x}{3} - 1 \leq \frac{1 - 2x}{7} + 4x$
17. $\frac{x}{a} - \frac{y}{b} \geq c$
18. $\frac{2x}{3a} + \frac{c}{2} \geq \frac{6y}{b}$
19. $x^3 - 6x^2 + 6 = 2$; find the root between 5 and 6 accurate to the nearest .1.
20. $6x^2 - x^3 + 2x + 1 = 0$; find the root between 6 and 7 accurate to the nearest .1.
21. $5x^3 - x - 5 = 0$; find the root between 1 and 2 accurate to the nearest .1.
22. $5x^4 + x^2 - 8 = 0$; find the root between 1 and 2 accurate to the nearest .1.

23. $3x - 1 > 2[2(7 - x)]$ and $2\left(\frac{x}{3} - \frac{6}{7}\right) < 5 - x$

24. $3 - x > 2$ or $-(3 - x) > 2$

Chapter 6 Review Answers

2. $x = 4; x \neq 2$
4. $x = 2$
6. x = (a↑2 + 7*a*b + b↑2)/(b - a); restriction $a \neq b \neq 0$.
8. $R = \dfrac{R_1R_2}{3R_2 + 2R_1}$; restriction $R_2 \neq -\dfrac{2}{3}R_1$
10. $x = -\frac{1}{3}$ or $x = \frac{7}{8}$
12. $x = 1.33$ or $x = -.93$
14. $x = 3.79$ or $x = -1.79$
16. $x \geq -.558$
18. $x < \dfrac{36ay - 3abc}{4b}$ if $a > 0$ and $x > \dfrac{36ay - 3abc}{4b}$ if $a < 0$
20. 6.3
22. 1.1
24. $x < 1$ or $x > 5$

Matrices

CHAPTER 7

7.0 INTRODUCTION

In the mathematics classroom today is often heard: What is the use of this? Sometimes there is no satisfactory concrete answer. A generation ago, the applications of matrix algebra would have been difficult to enumerate. With the advent of the computer, matrix notation is used extensively in problems of the business world.

Matrix algebra was first introduced by Arthur Cayley in 1858. For many years, the concept was a mathematical concept of interest only to pure mathematicians. In the 1930s matrix algebra was used in quantum mechanics, and it has subsequently been used in many physical and social sciences.

This chapter is designed to give the definitions and notations needed for matrix algebra with some problems to indicate how the concept is used in computer work. A second use of matrix algebra is to solve systems of linear equations, and this process will be explored in this chapter.

This chapter is designed to:

1. Provide the notation and define the operations of addition, subtraction, scalar multiplications, and generation of inverses of matrices.
2. Provide the procedure to generate the determinant value for any matrix.
3. Use matrix algebra to solve systems of linear equations.
4. Provide the notation and concepts needed to program computers to perform matrix algebra.

7.1 MATRIX ARITHMETIC

By definition a matrix is a rectangular array of numbers written as rows and columns enclosed in parentheses or brackets. Tables or lists can denote matrices if they are enclosed in parentheses or brackets. Lowercase letters represent the elements of a matrix, while capital letters represent the entire matrix.

$$A = \begin{pmatrix} 1 & 2 & 5 \\ 3 & 4 & 6 \end{pmatrix}$$

is a matrix with two rows and three columns. The dimension of A is 2×3. The elements are differentiated by subscripts that denote their locations within the matrix. Thus a_{21} is the designation of the element located in the second row, first column of matrix A; it is equal to 3. The general element of a matrix is denoted as a_{ij}, where i represents the row and j represents the column location.

$$B = \begin{pmatrix} b_{11} & b_{12} & b_{13} \\ b_{21} & b_{22} & b_{23} \end{pmatrix}$$

is a general matrix with the elements listed in general notation.

Equal matrices. Two matrices are *equal* if and only if the elements in the corresponding positions are equal and their dimensions are equal.

Using the matrices denoted by A and B, $A = B$ if and only if $1 = b_{11}$, $2 = b_{12}$, $5 = b_{13}$, $3 = b_{21}$, $4 = b_{22}$, and $6 = b_{23}$.

If

$$\begin{pmatrix} x+1 & 0 \\ 0 & y-3 \end{pmatrix} = \begin{pmatrix} 5 & 0 \\ 0 & -6 \end{pmatrix} \quad \text{then} \quad \begin{aligned} x + 1 &= 5 \quad \text{and} \quad y - 3 = -6 \\ x &= 4 \quad \text{and} \quad y = -3 \end{aligned}$$

Zero matrix. The *zero matrix* is a matrix in which all the elements are zero.

Square matrix. The *square matrix* is a matrix that has an equal number of rows and columns; the dimensions are $n \times n$.

Main diagonal. The *main diagonal* refers to the elements of a square matrix whose subscripts i and j are equal. The elements $a_{11}, a_{22}, a_{33}, \ldots, a_{nn}$ are the main diagonal of an $n \times n$ matrix A.

Unit matrix. The *unit matrix, or identity matrix*, is a square matrix with 1s on the main diagonal and 0s elsewhere.

$$\begin{pmatrix} 1 & 0 \\ 0 & 1 \end{pmatrix} \qquad 2 \times 2 \text{ identity matrix}$$

$$\begin{pmatrix} 1 & 0 & 0 \\ 0 & 1 & 0 \\ 0 & 0 & 1 \end{pmatrix} \qquad 3 \times 3 \text{ identity matrix}$$

Matrix Addition

To add two matrices, the dimensions must be equal. The sum of two matrices is the sum of the corresponding elements.

$$\begin{pmatrix} 1 & 2 & 3 \\ 4 & 5 & 6 \end{pmatrix}_{2\times 3} + \begin{pmatrix} -3 & 4 & -5 \\ 6 & -7 & 8 \end{pmatrix}_{2\times 3} = \begin{pmatrix} -2 & 6 & -2 \\ 10 & -2 & 14 \end{pmatrix}_{2\times 3}$$

$$\begin{pmatrix} a & c \\ b & d \end{pmatrix} + \begin{pmatrix} e & g \\ f & h \end{pmatrix} = \begin{pmatrix} a+e & c+g \\ b+f & d+h \end{pmatrix}$$

If

$$A = \begin{pmatrix} 1 & 2 \\ 3 & -4 \end{pmatrix} \quad \text{then} \quad -A = -\begin{pmatrix} 1 & 2 \\ 3 & -4 \end{pmatrix} = \begin{pmatrix} -1 & -2 \\ -3 & +4 \end{pmatrix}$$

Therefore,

$$\begin{pmatrix} 1 & 2 \\ 3 & 4 \end{pmatrix} - \begin{pmatrix} 4 & -3 \\ -2 & 1 \end{pmatrix} = \begin{pmatrix} -3 & 5 \\ 5 & 3 \end{pmatrix}$$

In matrix addition, $A + B = B + A$; that is, matrix addition is commutative. Though it may seem as if everything studied is commutative, a multiplication that is not commutative will soon be studied!

Matrix Multiplication

If a matrix is multiplied by a number (or a scalar), each element in the matrix is multiplied by the number.

Example

$$2\begin{pmatrix} a & c \\ b & d \end{pmatrix} = \begin{pmatrix} 2a & 2c \\ 2b & 2d \end{pmatrix}$$

If a matrix is multiplied by another matrix, several products and sums occur. First, those matrices that can be involved in a product need to be determined. The number of columns in the first matrix must equal the number of rows in the second matrix involved in the product. If this condition is not met, no product can be found. If the condition is met, the dimension of the product matrix is the (number of rows in the first matrix) × (the number of columns in the second matrix).

Example

A has dimensions 3×2, B has dimensions 2×2, and C has dimensions 2×3.

$A \cdot B$ can be found; the dimensions of AB are 3×2.
$B \cdot A$ cannot be found because $2 \neq 3$.
$A \cdot C$ can be found; the dimensions of AC are 3×3.
$C \cdot A$ can be found; the dimensions of CA are 2×2.
$B \cdot C$ can be found; the dimensions of BC are 2×3.
$C \cdot B$ cannot be found because $3 \neq 2$.

The results above certainly suggest that multiplication is not commutative in matrix multiplication, and that is the case: $AB \neq BA$.

After the dimensions have been analyzed and the dimensions of the resulting matrix determined, the individual elements must be found. In this case, $(2 \times 2) \cdot (2 \times 3)$ gives dimension 2×3.

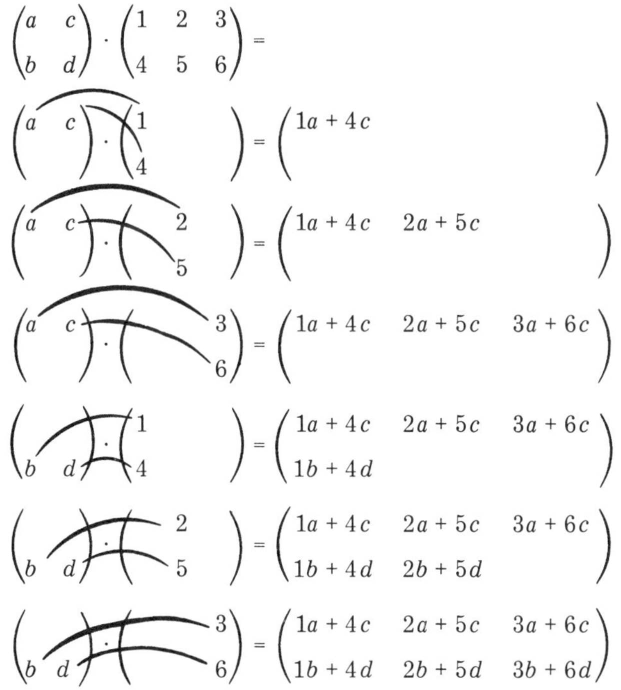

Row 1 of the first matrix must be multiplied by each column of the second matrix. The first row times the first column gives product a_{11}. The first row times the second column gives product a_{12}. The first row times the third column gives product a_{13}.

Row 2 of the first matrix multiplied by column 1 of the second matrix gives product a_{21}. Row 2 multiplied by column 2 gives product a_{22}. Row 2 multiplied by column 3 gives product a_{23}. The finished product is:

$$\begin{pmatrix} a & c \\ b & d \end{pmatrix}_{2\times 2} \cdot \begin{pmatrix} 1 & 2 & 3 \\ 4 & 5 & 6 \end{pmatrix}_{2\times 3} = \begin{pmatrix} a+4c & 2a+5c & 3a+6c \\ b+4d & 2b+5d & 3b+6d \end{pmatrix}_{2\times 3}$$

Procedure for Matrix Multiplication of *A* and *B* to Give *C*

1. The number of columns of A must equal the number of rows of B. The product, C, has dimensions equal to the number of rows of A by the number of columns of B.
2. The elements of row 1 of C are the result of multiplying the elements of row 1 of A by each of the columns of B.
3. The elements of row 2 of C are the result of multiplying the elements of row 2 of A by each of the columns of B.
4. Continue this procedure for the number of rows of A.

The identity matrix is very special in matrix multiplication. It acts very much like the identity element 1 in multiplication.

$$\begin{pmatrix} 1 & 0 & 0 \\ 0 & 1 & 0 \\ 0 & 0 & 1 \end{pmatrix} \cdot \begin{pmatrix} a_{11} & a_{12} & a_{13} \\ a_{21} & a_{22} & a_{23} \\ a_{31} & a_{32} & a_{33} \end{pmatrix} = \begin{pmatrix} a_{11} & a_{12} & a_{13} \\ a_{21} & a_{22} & a_{23} \\ a_{31} & a_{32} & a_{33} \end{pmatrix}$$

This product is commutative. The identity matrix can be multiplied on either side of a matrix (as long as the dimensions are legitimate) and the products are identical. I is used to designate the identity matrix, so the above statement can be written $AI = IA = A$.

Likewise, the zero matrix acts very much as the factor 0 in a product.

$$\begin{pmatrix} 0 & 0 \\ 0 & 0 \end{pmatrix} \cdot \begin{pmatrix} a & b & c \\ d & e & f \end{pmatrix} = \begin{pmatrix} 0 & 0 & 0 \\ 0 & 0 & 0 \end{pmatrix}$$

The operations of addition, subtraction, and multiplication of matrices have been discussed; the exercises give practice in the notations and procedures. The operation of division involves generating an inverse.

As an example of how matrices are used in the business world, suppose you operate three stores that handle televisions, stereos, and videocameras. The July 1 inventory for Store A was 3 televisions, 4 stereos, and 10 videocameras. The inventory for store B was 8 televisions, 5 stereos, and 2 videocameras; for store C the inventory was 6 televisions, 9 stereos, and 5 videocameras. The costs for these items were \$280 for television sets, \$400 for stereos, and \$700 for videocameras. The inventory value for each store is desired.

The inventory can be arranged as

$$\begin{array}{c} \\ A \\ B \\ C \end{array} \begin{array}{c} \begin{array}{ccc} TV & S & VC \end{array} \\ \begin{pmatrix} 3 & 4 & 10 \\ 8 & 5 & 2 \\ 6 & 9 & 5 \end{pmatrix} \end{array}$$

Because we have the prices for each article and a product is involved, the price matrix is

$$\begin{pmatrix} 280 \\ 400 \\ 700 \end{pmatrix}$$

The product is

$$\begin{array}{c} \\ A \\ B \\ C \end{array} \begin{array}{c} \begin{array}{ccc} TV & S & VC \end{array} \\ \begin{pmatrix} 3 & 4 & 10 \\ 8 & 5 & 2 \\ 6 & 9 & 5 \end{pmatrix}_{3 \times 3} \end{array} \cdot \begin{array}{c} TV \\ S \\ VC \end{array} \begin{pmatrix} 280 \\ 400 \\ 700 \end{pmatrix}_{3 \times 1} = \begin{array}{c} A \\ B \\ C \end{array} \begin{pmatrix} 11{,}240 \\ 7{,}880 \\ 8{,}780 \end{pmatrix}$$

The order of this multiplication problem could not be reversed, since a 3×3 matrix cannot be multiplied by a 3×1 matrix. To reverse the order, it would be necessary to write the price matrix as a 1×3 array.

EXERCISES

Given:

$$A = \begin{pmatrix} 1 & 2 \\ 3 & -1 \\ -2 & 3 \end{pmatrix}, \quad B = \begin{pmatrix} x & y & z \\ a & b & c \end{pmatrix}, \quad C = \begin{pmatrix} 5 \\ 10 \end{pmatrix},$$

$$D = (2 \quad 4 \quad 6), \quad E = \begin{pmatrix} 10 & b \\ a & -1 \\ 9 & 2 \end{pmatrix}, \quad F = \begin{pmatrix} 1 & 0 & 0 \\ 0 & 0 & 1 \\ 0 & 1 & 0 \end{pmatrix}$$

1. Give the dimensions of A, B, and C.
2. Give the dimensions of D, E, and F.
3. Identify a_{31}, b_{12}, and c_{21}.
4. Identify d_{13}, e_{21}, and f_{32}.
5. If $A = E$, find the value of a in matrix E.
6. If $A = E$, find the value of b in matrix E.
7. Find

$$A + \begin{pmatrix} 7 & -8 \\ 6 & 5 \\ 3 & 4 \end{pmatrix}$$

8. Find

$$B + \begin{pmatrix} 1 & 2 & 3 \\ 4 & 5 & 6 \end{pmatrix}$$

9. Find $A - E$.
10. Find $E - A$.
11. Find $6E$.
12. Find $4F$.
13. Find $A \cdot B$.
14. Find $B \cdot A$.
15. Find $C \cdot D$.
16. Find $D \cdot E$.
17. Find $(B \cdot E) \cdot F$.
18. Find $E \cdot (B \cdot F)$.
19. A company has three apartment complexes in a city. Complex A has 14 one-bedroom, 22 two-bedroom, and 8 three-bedroom apartments. Complex B has 20 one-bedroom, 14 two-bedroom, and 11 three-bedroom apartments. Complex C has 12 one-bedroom, 20 two-bedroom, and 12 three-bedroom apartments.

    Rentals gross \$260 a month for one bedroom, \$350 a month for two bedrooms, and \$420 a month for three bedrooms. Set up this information in matrix form and determine the maximum monthly income from each complex using matrix arithmetic.
20. A company has three types of automobiles for rental. The economy auto rents for \$15 per day, the standard model rents for \$20 a day, and the luxury model rents for \$30 a day. The company has businesses in four cities with the following inventory:

    City B has 52 standard, 60 economy, and 10 luxury cars.
    City C has 35 standard, 40 economy, and 8 luxury models.
    City D has 25 standard, 40 economy, and 5 luxury cars.
    City E has 40 standard, 50 economy and no luxury cars.

Set up this information in matrix form and determine the daily proceeds from each city if the company averages 80% rentals on a given day.

Answers

2. $D = 1 \times 3, E = 3 \times 2, F = 3 \times 3$
4. $d_{13} = 6, e_{21} = a, f_{32} = 1$
6. $b = 2$
8. $\begin{pmatrix} x+1 & y+2 & z+3 \\ a+4 & b+5 & c+6 \end{pmatrix}$
10. $\begin{pmatrix} 9 & b-2 \\ a-3 & 0 \\ 11 & -1 \end{pmatrix}$
12. $\begin{pmatrix} 4 & 0 & 0 \\ 0 & 0 & 4 \\ 0 & 4 & 0 \end{pmatrix}$
14. $\begin{pmatrix} x+3y-2z & 2x-y+3z \\ a+3b-2c & 2a-b+3c \end{pmatrix}$
16. $(74 + 4a \quad 8 + 2b)$
18. $\begin{pmatrix} 10x+ab & 10z+cb & 10y+b^2 \\ ax-a & az-c & ay-b \\ 9x+2b & 9z+2c & 9y+2b \end{pmatrix}$
20. $\begin{matrix} \text{B} \\ \text{C} \\ \text{D} \\ \text{E} \end{matrix} \begin{pmatrix} 1792 \\ 1232 \\ 1000 \\ 1240 \end{pmatrix}$

7.2 DETERMINANTS

Each square matrix has a real number associated with it called a *determinant.* The symbols used to denote a determinant are two vertical bars (| |).

The determinant of a 2×2 square matrix is $A = |A|$, where

$$|A| = \begin{vmatrix} a_{11} & a_{12} \\ a_{21} & a_{22} \end{vmatrix} = a_{11}a_{22} - a_{12}a_{21}$$

The value of the determinant is the product of the elements on the main diagonal minus the product of the elements on the minor diagonal.

The determinant associated with

$$\begin{pmatrix} 2 & -1 \\ 3 & 4 \end{pmatrix} \quad \text{is} \quad \begin{vmatrix} 2 & -1 \\ 3 & 4 \end{vmatrix} = (2)(4) - (-1)(3) = 11$$

Determinants are useful in solving systems of linear equations. Given two general linear equations

$$ax + by = c$$

$$dx + ey = f$$

By using linear combinations to solve these equations, the solution is determined to be

$$x = \frac{ce - bf}{ae - bd} \quad \text{and} \quad y = \frac{af - cd}{ae - bd}$$

The two denominators are identical and can be written as a determinant. The numerators take a bit more ingenuity, but the two solutions can be written in determinant form as

$$x = \frac{\begin{vmatrix} c & b \\ f & e \end{vmatrix}}{\begin{vmatrix} a & b \\ d & e \end{vmatrix}} = \frac{ce - bf}{ae - bd} \qquad y = \frac{\begin{vmatrix} a & c \\ d & f \end{vmatrix}}{\begin{vmatrix} a & b \\ d & e \end{vmatrix}} = \frac{af - cd}{ae - bd}$$

Referring back to the two general linear equations, the determinant of the denominator is the array of coefficients of x and y. When solving for x, the x-coefficients in the determinant of the numerator are replaced by the numerical terms on the right side of the equation; the y-coefficients are retained. In solving for y, the x-coefficients in the determinant of the numerator are retained; the y-coefficients are replaced by the numbers on the right side of the equation.

Example

Solve by determinants:

$$y = 3x - 5$$

$$x + 2y = -1$$

Before determinants can be used, the equations must be in general form.

$$3x - y = 5$$

$$x + 2y = -1$$

The determinant of the denominator is formed from the coefficients.

$$\begin{vmatrix} 3 & -1 \\ 1 & 2 \end{vmatrix} = (3)(2) - (-1)(1) = 7$$

The determinant of the numerator for x is formed from the number terms 5 and -1 and the coefficients of y.

$$\begin{vmatrix} 5 & -1 \\ -1 & 2 \end{vmatrix} = (5)(2) - (-1)(-1) = 9$$

The determinant of the numerator for y is formed from the number terms 5 and -1 and the coefficients of x.

$$\begin{vmatrix} 3 & 5 \\ 1 & -1 \end{vmatrix} = (3)(-1) - (5)(1) = -8$$

Therefore, the solution to the system of equations is $x = \frac{9}{7}$ and $y = -\frac{8}{7}$. There are two special cases to consider in solving a linear system of equations by determinants. If the determinant value of the numerator and denominator are both equal to 0, there is no unique solution to the problem; the system is called *dependent.* If only the denominator is equal to 0, there is no solution to the system and the system is called *inconsistent.*

3 X 3 Determinant

In order to find the determinant value for a 3 X 3 matrix, it is necessary to define a minor and a cofactor.

Given a 3 X 3 matrix

$$A = \begin{pmatrix} a_{11} & a_{12} & a_{13} \\ a_{21} & a_{22} & a_{23} \\ a_{31} & a_{32} & a_{33} \end{pmatrix}$$

The minor of an element a_{ij} in a determinant is the determinant that remains after the ith row and the jth column have been deleted.

The minor of a_{21} is

$$\begin{vmatrix} a_{12} & a_{13} \\ a_{32} & a_{33} \end{vmatrix}$$

The minor of a_{12} is

$$\begin{vmatrix} a_{21} & a_{23} \\ a_{31} & a_{33} \end{vmatrix}$$

The cofactor of the ijth element in a square matrix is

$$(-1)^{i+j} \cdot \text{minor of } a_{ij}$$

Given

$$\begin{pmatrix} 1 & -2 & 5 \\ -3 & 4 & 3 \\ 2 & 6 & -4 \end{pmatrix}$$

The cofactor of a_{31}, or 2, is

$$(-1)^{3+1} \begin{vmatrix} -2 & 5 \\ 4 & 3 \end{vmatrix} = (+1)\,(-6 - 20) = -26$$

A determinant of a 3 $\times$ 3 matrix is equal to the sum of the products formed by multiplying the entries of any row or column and their cofactors. If we wish to find the determinant of the above matrix, we may choose any row or column. Choose row 1:1 -2 5, and determine the cofactors of each of these.

The cofactor of 1 is $(-1)^{1+1} \begin{vmatrix} 4 & 3 \\ 6 & -4 \end{vmatrix} = (1)\,(-16 - 18) = -34.$

The cofactor of -2 is $(-1)^{1+2} \begin{vmatrix} -3 & 3 \\ 2 & -4 \end{vmatrix} = (-1)\,(12 - 6) = -6.$

The cofactor of 5 is $(-1)^{1+3} \begin{vmatrix} -3 & 4 \\ 2 & 6 \end{vmatrix} = (1)\,(-18 - 8) = -26.$

The definition for the value of the determinant of a 3 $\times$ 3 matrix is the entry times its cofactor. The determinant value is $(1)\,(-34) + (-2)\,(-6) + (5)\,(-26) = -34 + 12 - 130 = -152$.

Any other row or column may be chosen to calculate the determinant; the same numerical answer results. If any zeros exist in a matrix, the choice of the row or column containing the zero(s) simplifies the calculations.

Determinants of 3 $\times$ 3 matrices are useful in solving simultaneous equations of three equations and three unknowns. The same format as for two equations and two unknowns is used.

Example

Solve

$$x + 2y + 3z = 26$$
$$3x + 5y + 2z = 39$$
$$2x + 4y + z = 27$$

by determinants.

$$x = \frac{\begin{vmatrix} 26 & 2 & 3 \\ 39 & 5 & 2 \\ 27 & 4 & 1 \end{vmatrix}}{\begin{vmatrix} 1 & 2 & 3 \\ 3 & 5 & 2 \\ 2 & 4 & 1 \end{vmatrix}} \qquad y = \frac{\begin{vmatrix} 1 & 26 & 3 \\ 3 & 39 & 2 \\ 2 & 27 & 1 \end{vmatrix}}{\text{(Denominator)}} \qquad z = \frac{\begin{vmatrix} 1 & 2 & 26 \\ 3 & 5 & 39 \\ 2 & 4 & 27 \end{vmatrix}}{\text{(Denominator)}}$$

The denominator is $(1)(-3) + (-3)(-10) + (2)(-11) = 5$, using column 1.
The numerator of x, using column 1, is $26(-3) + (-39)(-10) + (27)(-11) = 15$.
The numerator of y, using column 2, is $(-26)(-1) + (39)(-5) + (-27)(-7) = 20$.

The value of x is $\frac{15}{5} = 3$. The value of y is $\frac{20}{5} = 4$. Substitute these values in one of the original equations; the value of z is 5 or solve for z using determinants.

EXERCISES

1. Determine the determinant of $\begin{pmatrix} 3 & 4 \\ -1 & 5 \end{pmatrix}$.
2. Determine the determinant of $\begin{pmatrix} 7 & 4 \\ 5 & -3 \end{pmatrix}$.
3. Give a 2×2 matrix that has a determinant of $ab - cd$.
4. Give a 2×2 matrix that has a determinant of $3x + 2y$.

Given

$$A = \begin{pmatrix} 6 & 1 & 2 \\ 4 & -3 & 0 \\ 5 & 7 & 8 \end{pmatrix}$$

5. Find the minor of 0.
6. Find the minor of 1.
7. Find the cofactor of 4.
8. Find the cofactor of 7.
9. Find the $|A|$ using row 1.
10. Find the $|A|$ using column 3.
11. Solve the following system of linear equations by determinants.

$$5x + y + 1 = 0$$

$$2y = 1 - x$$

12. Solve the following system of linear equations by determinants.

$$8y + x = 2$$
$$2x - 5y = 1$$

13. Solve the following system of linear equations by determinants.

$$x + y - 3z = 2$$
$$y + 3z + 1 = 0$$
$$x + 2y - 1 = 0$$

14. Solve the following system of linear equations by determinants.

$$x + 4y - 2z = 1$$
$$2y + 3z = 0$$
$$x + 8y + 2z = 0$$

15. Solve the following system by determinants.

$$4x - 2y = 7$$
$$2x - y = 7$$

16. Solve the following system of linear equations by determinants.

$$3x - 2y = 5$$
$$4y - 6x + 10 = 0$$

Answers

2. -41
4. $\begin{vmatrix} 3 & -2 \\ y & x \end{vmatrix}$, $\begin{vmatrix} x & 2 \\ -y & 3 \end{vmatrix}$ or $\begin{vmatrix} x & y \\ -2 & 3 \end{vmatrix}$
6. $\begin{vmatrix} 4 & 0 \\ 5 & 8 \end{vmatrix} = 32$
8. 8
10. -90
12. $x = \frac{6}{7}, y = \frac{1}{7}$
14. $x = 5, y = -\frac{3}{4}, z = \frac{1}{2}$
16. Dependent

7.3 INVERSE MATRICES

The *inverse matrix*, of A, denoted by A^{-1} is related to a square matrix A in such a way that $A^{-1} \cdot A = I$ or $A \cdot A^{-1} = I$ whenever $|A| \neq 0$. Recall that I is the identity matrix.

Example

Given

$$A = \begin{pmatrix} 2 & 5 \\ -1 & 1 \end{pmatrix}, \quad \text{and} \quad A^{-1} = \frac{1}{7}\begin{pmatrix} 1 & -5 \\ 1 & 2 \end{pmatrix} = \begin{pmatrix} \frac{1}{7} & -\frac{5}{7} \\ \frac{1}{7} & \frac{2}{7} \end{pmatrix}.$$

$$A^{-1} \cdot A = \begin{pmatrix} \frac{1}{7} & -\frac{5}{7} \\ \frac{1}{7} & \frac{2}{7} \end{pmatrix} \cdot \begin{pmatrix} 2 & 5 \\ -1 & 1 \end{pmatrix} = \begin{pmatrix} 1 & 0 \\ 0 & 1 \end{pmatrix}$$

$$A \cdot A^{-1} = \begin{pmatrix} 2 & 5 \\ -1 & 1 \end{pmatrix} \cdot \begin{pmatrix} \frac{1}{7} & -\frac{5}{7} \\ \frac{1}{7} & \frac{2}{7} \end{pmatrix} = \begin{pmatrix} 1 & 0 \\ 0 & 1 \end{pmatrix}$$

Procedure for Determining the Inverse of a 2 × 2 Matrix A

1. Find the value of $|A|$. If $|A| = 0$, no inverse exists.
2. Interchange elements on the main diagonal.
3. Multiply the elements on the minor diagonal by -1.
4. Multiply the resulting matrix by $1/|A|$.

Example

$$B = \begin{pmatrix} 3 & -1 \\ 1 & 2 \end{pmatrix}. \quad \text{Find } B^{-1}.$$

Step 1. $|B| = 6 - (-1) = 7$

Step 2. $\begin{pmatrix} 2 & \\ & 3 \end{pmatrix}$

Step 3. $\begin{pmatrix} 2 & +1 \\ -1 & 3 \end{pmatrix}$

Step 4. $\frac{1}{7}\begin{pmatrix} 2 & +1 \\ -1 & 3 \end{pmatrix} = \begin{pmatrix} \frac{2}{7} & \frac{1}{7} \\ -\frac{1}{7} & \frac{3}{7} \end{pmatrix} = B^{-1}$

Check: $\begin{pmatrix} \frac{2}{7} & \frac{1}{7} \\ -\frac{1}{7} & \frac{3}{7} \end{pmatrix} \begin{pmatrix} 3 & -1 \\ 1 & 2 \end{pmatrix} = \begin{pmatrix} 1 & 0 \\ 0 & 1 \end{pmatrix}$

Inverse matrices are used in the solution of systems of linear equations.

Example

$$3x - y = 5$$
$$x + 2y = -1$$

This system can be written in matrix multiplication notation as:

$$\begin{pmatrix} 3 & -1 \\ 1 & 2 \end{pmatrix} \cdot \begin{pmatrix} x \\ y \end{pmatrix} = \begin{pmatrix} 5 \\ -1 \end{pmatrix}$$

To solve for x and y, multiply both sides of the matrix equation by the inverse of

$$\begin{pmatrix} 3 & -1 \\ 1 & 2 \end{pmatrix}$$

Be sure to put the inverse on the left of the existing matrices on each side; otherwise, multiplication cannot be performed.

$$\begin{pmatrix} \frac{2}{7} & \frac{1}{7} \\ -\frac{1}{7} & \frac{3}{7} \end{pmatrix} \cdot \begin{pmatrix} 3 & -1 \\ 1 & 2 \end{pmatrix} \begin{pmatrix} x \\ y \end{pmatrix} = \begin{pmatrix} \frac{2}{7} & \frac{1}{7} \\ -\frac{1}{7} & \frac{3}{7} \end{pmatrix} \cdot \begin{pmatrix} 5 \\ -1 \end{pmatrix}$$

$$\begin{pmatrix} 1 & 0 \\ 0 & 1 \end{pmatrix} \cdot \begin{pmatrix} x \\ y \end{pmatrix} = \begin{pmatrix} \frac{9}{7} \\ -\frac{8}{7} \end{pmatrix}$$

$$\begin{pmatrix} x \\ y \end{pmatrix} = \begin{pmatrix} \frac{9}{7} \\ -\frac{8}{7} \end{pmatrix}$$

Therefore $x = \frac{9}{7}$ and $y = -\frac{8}{7}$.

Procedure for Determining the Inverse of a 3 × 3 Square Matrix A

1. Find $|A|$. If $|A| = 0$, no inverse exists.
2. Find the cofactors of each element of A and replace the element by its cofactor.
3. Transpose the new matrix (write row 1 as column 1, row 2 as column 2, and so on).
4. Multiply the transpose by $1/|A|$.

Example

$$\text{Given matrix } A = \begin{pmatrix} 1 & 2 & 3 \\ 3 & 5 & 2 \\ 2 & 4 & 1 \end{pmatrix}$$

Step 1. $|A| = 1\,(-3) + (-3)\,(-10) + (2)\,(-11) = 5$

Step 2. $\begin{pmatrix} -3 & +1 & 2 \\ +10 & -5 & 0 \\ -11 & +7 & -1 \end{pmatrix}$ Cofactors of each element

Step 3. $\begin{pmatrix} -3 & 10 & -11 \\ 1 & -5 & 7 \\ 2 & 0 & -1 \end{pmatrix}$ Transpose of cofactors

Step 4. $A^{-1} = \begin{pmatrix} -\frac{3}{5} & 2 & -\frac{11}{5} \\ \frac{1}{5} & -1 & \frac{7}{5} \\ \frac{2}{5} & 0 & -\frac{1}{5} \end{pmatrix}$ Product with $\frac{1}{|A|}$

Check:

$$\begin{pmatrix} -\frac{3}{5} & 2 & -\frac{11}{5} \\ \frac{1}{5} & -1 & \frac{7}{5} \\ \frac{2}{5} & 0 & -\frac{1}{5} \end{pmatrix} \cdot \begin{pmatrix} 1 & 2 & 3 \\ 3 & 5 & 2 \\ 2 & 4 & 1 \end{pmatrix} = \begin{pmatrix} 1 & 0 & 0 \\ 0 & 1 & 0 \\ 0 & 0 & 1 \end{pmatrix}$$

Inverse matrices can be used in solving a system of three linear equations in three unknowns.

Example

Given the system:

$$x + 2y + 3z = 26$$
$$3x + 5y + 2z = 39$$
$$2x + 4y + z = 27$$

This system can be written in matrix notation as:

$$\begin{pmatrix} 1 & 2 & 3 \\ 3 & 5 & 2 \\ 2 & 4 & 1 \end{pmatrix} \cdot \begin{pmatrix} x \\ y \\ z \end{pmatrix} = \begin{pmatrix} 26 \\ 39 \\ 27 \end{pmatrix}$$

Multiply both sides of this matrix equation by the inverse of the matrix on left. The inverse was generated in the example above.

$$\begin{pmatrix} -\frac{3}{5} & 2 & -\frac{11}{5} \\ \frac{1}{5} & -1 & \frac{7}{5} \\ \frac{2}{5} & 0 & -\frac{1}{5} \end{pmatrix} \cdot \begin{pmatrix} 1 & 2 & 3 \\ 3 & 5 & 2 \\ 2 & 4 & 1 \end{pmatrix} \begin{pmatrix} x \\ y \\ z \end{pmatrix} = \begin{pmatrix} -\frac{3}{5} & 2 & -\frac{1}{5} \\ \frac{1}{5} & -1 & \frac{7}{5} \\ \frac{2}{5} & 0 & -\frac{1}{5} \end{pmatrix} \cdot \begin{pmatrix} 26 \\ 39 \\ 27 \end{pmatrix}$$

$$\begin{pmatrix} 1 & 0 & 0 \\ 0 & 1 & 0 \\ 0 & 0 & 1 \end{pmatrix} \cdot \begin{pmatrix} x \\ y \\ z \end{pmatrix} = \begin{pmatrix} 3 \\ 4 \\ 5 \end{pmatrix}$$

$$\begin{pmatrix} x \\ y \\ z \end{pmatrix} = \begin{pmatrix} 3 \\ 4 \\ 5 \end{pmatrix}$$

This gives the solution $x = 3$, $y = 4$, and $z = 5$.

The same procedure can be used to find the inverse of a large matrix. It is obviously quite a complex and lengthy procedure. In scientific work, it is often necessary to find the inverse of a matrix with hundreds of rows. Several different procedures have been developed to perform this computation. The statements for matrix operations available in computer work are among the most powerful and useful in the system and have made matrix algebra a useful tool.

EXERCISES

1. Given $A = \begin{pmatrix} 5 & 3 \\ 7 & 4 \end{pmatrix}$, find A^{-1}.

2. Given $A = \begin{pmatrix} 3 & 2 \\ 5 & 4 \end{pmatrix}$, find A^{-1}.

3. Given $A = \begin{pmatrix} 7 & 1 \\ -3 & 1 \end{pmatrix}$, find A^{-1}.

4. Given $A = \begin{pmatrix} 5 & -2 \\ -8 & 4 \end{pmatrix}$, find A^{-1}.

5. Given $A = \begin{pmatrix} 1 & 1 \\ 1 & 1 \end{pmatrix}$, find A^{-1}.

6. Given $A = \begin{pmatrix} -2 & -2 \\ -2 & -2 \end{pmatrix}$, find A^{-1}.

7. Solve the system

$$5x + 3y = 11$$
$$7x + 4y = 20$$

by matrix algebra. Use the results of Exercise 1.

8. Solve the system

$$3x + 2y = 41$$
$$5x + 4y = 37$$

by matrix algebra. Use the results of Exercise 2.

9. Solve the system

$$7x + y = 1.06$$
$$-3x + y = 3.2$$

by matrix algebra. Use the results of Exercise 3.

10. Solve the system

$$5x - 2y = 2.8$$
$$-8x + 4y = 5.1$$

by matrix algebra. Use the results of Exercise 4.

11. Given $A = \begin{pmatrix} 1 & 0 & 3 \\ 1 & 1 & 3 \\ 0 & 2 & 2 \end{pmatrix}$, find A^{-1}.

12. Given $A = \begin{pmatrix} 4 & 4 & -6 \\ -4 & -6 & 6 \\ 2 & 4 & -4 \end{pmatrix}$, find A^{-1}.

13. Given $A = \begin{pmatrix} 1 & 1 & -2 \\ 2 & 0 & 1 \\ 1 & -1 & 1 \end{pmatrix}$, find A^{-1}.

14. Given $A = \begin{pmatrix} 3 & -1 & 2 \\ 1 & 1 & 0 \\ 2 & 0 & 1 \end{pmatrix}$, find A^{-1}.

15. Solve for x, y, and z (see Exercise 11):

$$x + 3z = 10$$
$$x + y + 3z = 8$$
$$2y + 2z = 6$$

16. Solve for x, y, and z (see Exercise 12):

$$4x + 4y - 6z = 23$$
$$-4x - 6y + 6z = 17$$
$$2x + 4y - 4z = 20$$

17. Solve for x, y, and z (see Exercise 13):

$$x + y - 2z = 1.8$$
$$2x + z = 2.1$$
$$x - y + z = 1.7$$

18. Solve for x, y, and z (see Exercise 14):

$$\begin{aligned} 3x - y + 2z &= 7.1 \\ x + y &= 5.2 \\ 2x + z &= 6.0 \end{aligned}$$

19. A manufacturing company uses three types of machines, A, B, and C, in its plant. Each machine produces three products, x, y, and z, per shift. Total production per machine per shift is shown in the following table:

	Machine		
Product	A	B	C
x	2	2	2
y	0	3	6
z	2	1	1

How many machines of each type are in use in the plant if the total production per shift is 18 units of product x, 30 units of product y, and 12 units of product z?

20. Repeat Exercise 19, except 20 units of product x, 27 units of product y, and 15 units of product z are needed at the end of the shift.

Answers

2. $|A| = 2, \quad A^{-1} = \begin{pmatrix} 2 & -1 \\ -\frac{5}{2} & \frac{3}{2} \end{pmatrix}$

4. $|A| = 4, \quad A^{-1} = \begin{pmatrix} 1 & \frac{1}{2} \\ 2 & \frac{5}{4} \end{pmatrix}$

6. $|A| = 0$; no inverse exists.

8. $x = 45, \quad y = -47$

10. $x = 5.35, \quad y = 11.975$

12. $|A| = 8, \quad A^{-1} = \begin{pmatrix} 0 & -1 & -\frac{3}{2} \\ -\frac{1}{2} & -\frac{1}{2} & 0 \\ -\frac{1}{2} & -1 & -1 \end{pmatrix}$

14. $A = 0$; no inverse exists.

16. $x = -47, \quad y = -20, \quad z = -48.5$

18. No solution

20. 5 machine A, 1 machine B, and 4 machine C

7.4 MATRIX MULTIPLICATION ON A COMPUTER

Once the operations in matrix arithmetic are defined, the actual process involves only the most basic arithmetic processes—addition, subtraction, and multiplication. However, the number of these processes makes the work tedious and prone to error. The computer does these processes rapidly and accurately. Programming these operations involves some clever use of matrix notation.

In general notation, the product of A and B, or C, is as follows:

$$\begin{pmatrix} a_{11} & a_{12} & a_{13} \\ a_{21} & a_{22} & a_{23} \\ a_{31} & a_{32} & a_{33} \end{pmatrix} \cdot \begin{pmatrix} b_{11} & b_{12} & b_{13} \\ b_{21} & b_{22} & b_{23} \\ b_{31} & b_{32} & b_{33} \end{pmatrix} = \begin{pmatrix} c_{11} & c_{12} & c_{13} \\ c_{21} & c_{22} & c_{23} \\ c_{31} & c_{32} & c_{33} \end{pmatrix}$$

where

$$c_{11} = a_{11}b_{11} + a_{12}b_{21} + a_{13}b_{31}$$

$$c_{23} = a_{21}b_{13} + a_{22}b_{23} + a_{23}b_{33}$$

A pattern exists in these products.

$$c_{ij} = a_{i1}b_{1j} + a_{i2}b_{2j} + a_{i3}b_{3j}$$

For any entry of the product matrix, a row of the first matrix and a column of the second matrix are involved.

To generalize this, it is necessary to introduce a notation that denotes a sum. If $S = a_1 + a_2 + a_3 + a_4 + a_5$, the mathematical notation to denote this is

$$S = \sum_{n=1}^{5} a_n$$

The Greek letter sigma (Σ) is used to denote a sum. The notation is read as the sum of a_n as n changes from 1 to 5. This notation refers to integer values of n.

$$\sum_{i=2}^{6} a_i = a_2 + a_3 + a_4 + a_5 + a_6$$

In the expression $c_{11} = a_{11}b_{11} + a_{12}b_{21} + a_{13}b_{31}$, the first subscript is constant throughout each term, the second takes on different values from 1 to 3, the third takes on different values from 1 to 3 (but the same values as the second), and the last is constant throughout each term. Thus $c_{11} = a_{1k}\,b_{k1}$ as k takes on values 1, 2, and 3 in successive terms. This can be written in sigma notation as

$$c_{11} = \sum_{k=1}^{3} a_{1k} b_{k1}$$

Using the above notation, we should be able to write any term in the product.

$$c_{③②} = \sum_{k=1}^{3} a_{③k} b_{k②}$$

Therefore, generalizing even more, it is possible to write any element in a product. The most general term in a matrix product is c_{ij}.

$$c_{ij} = \sum_{k=1}^{n} a_{ik} b_{kj}$$

Using this notation, a flowchart can be written for the product of two matrices.

Flowchart of a Matrix Product of Matrix $A_{m \times t}$ and Matrix $B_{t \times n}$

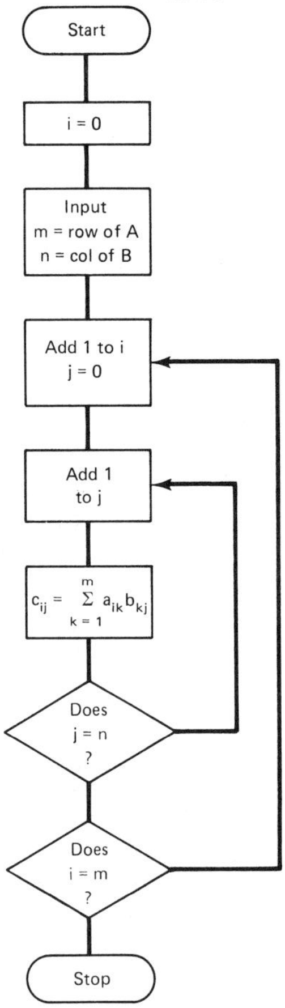

EXERCISES

1. If A has dimension 3×4 and B has dimension 4×2, find c_{21}, where $C = A \cdot B$. Write c_{21} in both expanded notation and sigma notation.
2. If A has dimension 4×4 and B has dimension 4×2, find c_{32}, where $C = A \cdot B$. Write c_{32} in both expanded form and sigma notation.
3. If A has dimension 25×25 and B has dimension 25×30, find $c_{7,21}$, where $C = A \cdot B$. Write $c_{7,21}$ in sigma notation.
4. If A has dimension 25×30 and B has dimension 30×20, find $c_{19,8}$, where $C = A \cdot B$. Write $c_{19,8}$ in sigma notation.
5. If

$$A = \begin{pmatrix} -3 & 3 & -3 & 2 \\ 3 & -4 & 4 & -2 \\ -3 & 4 & -5 & 3 \\ 2 & -2 & 3 & -2 \end{pmatrix}, \quad \text{then} \quad A^{-1} = \begin{pmatrix} 0 & 1 & 2 & 2 \\ 1 & 1 & 2 & 3 \\ 2 & 2 & 2 & 3 \\ 2 & 3 & 3 & 3 \end{pmatrix}$$

Solve the following system for y.

$$\begin{aligned} -3w + 3x - 3y + 2z &= -11 \\ 3w - 4x + 4y - 2z &= 13 \\ -3w + 4x - 5y + 3z &= -15 \\ 2w - 2x + 3y - 2z &= 9 \end{aligned}$$

Show sigma notation, expanded form, and the solution.
6. Use the information in Exercise 5. Solve for w. Show sigma notation, expanded form, and the solution.
7. Write the general form for finding c_{ij} if $C = A + B$ and A is a matrix of dimension 25×20 and B is a matrix of dimension 25×20.
8. Write the general form for finding c_{ij} if $C = A + B$ and A is a matrix of dimension 35×20 and B is a matrix of dimension 35×20.

Answers

2. $c_{32} = a_{31}b_{12} + a_{32}b_{22} + a_{33}b_{32} + a_{34}b_{42}$; $\quad c_{32} = \sum_{k=1}^{4} a_{3k}b_{k2}$

4. $c_{19,8} = \sum_{k=1}^{30} a_{19,k}b_{k,20}$

6. $w = (0)(-11) + (1)(13) + (2)(-15) + (2)(9)$; $\quad w = 1$;

$$w = \sum_{k=1}^{4} a_{1k}n_{k1}$$

where n refers to the number matrix

$$\begin{pmatrix} -11 \\ 13 \\ -15 \\ 9 \end{pmatrix}$$

8. $c_{ij} = a_{ij} + b_{ij}$

CHAPTER 7 REVIEW EXERCISES

$$A = \begin{pmatrix} 3 & -1 & 7 \\ 5 & 2 & -4 \\ 0 & 6 & 9 \end{pmatrix}, \quad B = \begin{pmatrix} -6 & 5 \\ 1 & 4 \\ -9 & -3 \end{pmatrix}, \quad C = \begin{pmatrix} 0 & 2 \\ 2 & 0 \end{pmatrix}$$

$$D = \begin{pmatrix} 2 & 7 & -1 \\ -3 & 5 & 16 \end{pmatrix}, \quad E = \begin{pmatrix} x+1 & a \\ 1 & 4 \\ y-7 & b \end{pmatrix}, \quad F = \begin{pmatrix} 1 & 2 \\ 3 & 4 \end{pmatrix}$$

1. Give the dimensions of A, B, and C.
2. Give the dimensions of D, E, and F.
3. Solve for x and a in matrix E if $E = B$.
4. Solve for y and b in matrix E if $E = B$.
5. Determine the products that are possible with matrices A, B, and C.
6. Determine the products that are possible with matrices D, E, and F.
7. Find $B + E$.
8. Find $C + F$.
9. Find F^{-1}.
10. Find C^{-1}.
11. Solve the system:

$$x + 2y = 11$$
$$3x + 4y = -1$$

12. Solve the system:

$$1.1x - 3.2y = .03$$
$$2.7x + .8y = 2.9$$

13. Determine $(A \cdot B)_{22}$. Write in sigma notation and in expanded notation.
14. Determine $(E \cdot F)_{31}$. Write in sigma notation and in expanded notation.
15. Determine $|A|$.
16. Find the cofactor of a_{21}.

17. Determine the inverse of A.

18. Solve the system:

$$3x - y + 7z = 12$$
$$5x + 2y - 4z = 23$$
$$0 + 6y + 9z = 17$$

Chapter 7 Review Answers

2. The dimension of D is 2×3.
   The dimension of E is 3×2.
   The dimension of F is 2×2.

4. $y - 7 = -9, \quad y = -2; \quad b = -3$

6. $D \cdot E, \quad F \cdot D, \quad E \cdot F$

8. $C + F = \begin{pmatrix} 1 & 4 \\ 5 & 4 \end{pmatrix}$

10. $C^{-1} = \begin{pmatrix} 0 & \frac{1}{2} \\ \frac{1}{2} & 0 \end{pmatrix}$

12. $x = .977, \quad y = .327$

14. $\sum_{k=1}^{2} e_{3k} f_{k1} = (y - 7)(1) + (3)(b) = y - 7 + 3b$

16. 51

18. $x = 3.96, \quad y = 2.31, \quad z = .35$

Mathematical Sequences and Series

CHAPTER 8

8.0 INTRODUCTION

Observations in real-world situations sometimes produce numbers that follow a particular pattern. Mathematicians and scientists study and generalize these patterns, producing rules.

Knowing patterns can put order into events. Imagine finding that the height of the tides follows a pattern of (approximately) 28 days. It then became possible to tie the height of tides to the lunar orbit, and many mysteries of nature were solved.

In this chapter, some of the more common sequences are studied, general relationships are determined, and applications of these sequences to the world of business and science are discussed.

This chapter is designed to:

1. Provide the notation of sequences and series.
2. Provide the formulas for determining terms and sums of arithmetic and geometric sequences.
3. Provide applications of sequences and series in the business world.

8.1 SEQUENCES

A *sequence* is a functional relationship between a set of numbers and the set of natural numbers. The word *functional* can be replaced by the word *unique*. Natural numbers are the counting numbers.

Notation for sequences utilizes subscripts to denote the location of a term in a sequence; the letter a represents an element of the sequence, and the subscript following a denotes the term. Thus a_1 refers to the first term of a sequence, a_2 refers to the second term, and a_n is the notation for any term (referred to as the nth term) in the sequence. At least three terms are needed to determine a sequence; the relationship that exists among the three elements must continue throughout the sequence.

A variety of information can be determined from a sequence. Given three terms that form a pattern, subsequent terms can be generated, general relationships can be determined, and any term can be predicted without generating all of the terms preceding it.

A few concrete examples will help to demonstrate this information.

Example

1, 2, 3, . . . : This sequence is the counting numbers, and the next elements are 4, 5, ... To determine their relationship to the counting numbers, the elements are associated with the counting numbers.

> The counting number 1, the first element 1 (abbreviated notation (1, 1))
> The counting number 2, the second element 2, (2, 2)
> The counting number 3, the third element 3, (3, 3)

The element is the counting number, so the relationship is $a_n = n$. This relationship can be stated as "any element of the sequence (a_n) is equal to its position in the sequence." That is, $a_{20} = 20$, $a_{142} = 142$, and so on.

Example

Find a_{20} in the sequence 3, 5, 7, ...

The three dots at the end of the sequence indicate the sequence continues on in this pattern without end. The sequence 3, 5, 7, . . . , 19 ends at the number 19. The relationship between the counting numbers and the elements of the sequence can most easily be explored by associating the two as the ordered pairs (1, 3), (2, 5), and (3, 7). If we look at only the first pair, 3 is 3×1, 3 is $2 + 1$, and so on. Both of these ideas can be eliminated because the relationship in the second set does not fit; $5 \neq 3 \times 2$ and $5 \neq 2 + 2$. The relationship that holds throughout the sequence is $a_n = 2n + 1$. Using this relationship, it is not necessary to generate all the terms to determine a_{20}. Instead, substitute 20 into $a_n = 2n + 1$: $a_{20} = 2(20) + 1 = 41$.

Example

Find a_{20} in the sequence 2, 4, 8 . . . .

The relationship between the counting numbers and the elements of the sequence gives (1, 2), (2, 4), (3, 8) . . . . Hasty determination of the relationship might yield the formula $a_n = 2n$. However, the $n = 3$ does not produce 8 with this relationship; therefore, it is incorrect. The numbers 2, 4, 8 are all powers of 2; that is, $2 = 2^1$, $4 = 2^2$, and $8 = 2^3$. Therefore, the correct relationship is $a_n = 2^n$. The 20th term is $2^{20} = 1{,}048{,}576$.

Sometimes it is impossible to find a relationship between the elements and the counting numbers. Instead the relationship must be described in terms of the elements preceding the required term. When this kind of relationship is used, the rule is called a *recursive rule.*

Example

Find a rule for 1, 2, 3, 5, 8 . . . .

If we look only at the first three terms, the rule appears to be $a_n = n$; however, the fourth and fifth terms do not follow that pattern. After the first two terms, it appears that any successive term is the sum of the two preceding terms. This rule can only be effective after the first two terms have been defined. The rule would be written: $a_1 = 1$, $a_2 = 2$, and $a_{n+2} = a_n + a_{n+1}$. This formula indicates that the first term generated by the general rule is the third term: $a_{1+2} = a_1 + a_{1+1}$, or $a_3 = a_1 + a_2 = 1 + 2 = 3$.

EXERCISES

Given the general notation, generate the first five terms of each sequence.

1. $a_n = 2n + 1$
2. $a_n = 3n + 2$
3. $a_n = 5 - 3n$
4. $a_n = 3 - 2n$
5. $a_n = n$
6. $a_n = 2n$
7. $a_n = 2n - 1$
8. $a_n = (-1)^n (3n - 2)$
9. $a_n = (-1)^{n+1} (2^n)$
10. $a_n = 2^{n-1}$
11. $a_n = \frac{10}{n}$
12. $a_n = \frac{25}{n}$
13. $a_{n+2} = a_{n+1} + a_n$ if $a_1 = 1, a_2 = 1$
14. $a_{n+2} = 2a_{n+1} + a_n$ if $a_1 = 1, a_2 = 1$
15. $a_n = n^2$
16. $a_n = (n - 1)^2$
17. $a_n = \frac{n+1}{n}$

18. $a_n = \dfrac{2n - 1}{n + 1}$

19. $a_n = \dfrac{1}{n}$

20. $a_n = \dfrac{1}{(n + 1)}$

Find the general relationships for the following sequences.

21. 4, 5, 6, . . .
22. 5, 7, 9, . . .
23. 1, 2, 4, 8, . . .
24. $1, \frac{1}{2}, \frac{1}{3}, \ldots$
25. .1, .01, .001, . . .
26. .9, .99, .999, . . .
27. 2, 3, 5, 8, 13, . . .
28. $\frac{1}{2}, \frac{1}{2}, 1, \frac{3}{2}, \frac{5}{2}, \ldots$
29. Find the first eight terms of the following sequences. Round to the nearest tenth in each calculation.

$$a_{n+1} = \frac{16/a_n + a_n}{2}, \qquad a_1 = 1$$

30. Find the first eight terms of the following sequence. Round to the nearest tenth in each calculation.

$$a_{n+1} = \frac{16/a_n + a_n}{2}, \qquad a_1 = 2$$

31. How are the results in Exercises 29 and 30 related to the constant 16?
32. Predict the general recursive formula for $\sqrt{30}$.

Answers

2. 5, 8, 11, 14, 17
4. 1, -1, -3, -5, -7
6. 2, 4, 6, 8, 10
8. -1, 4, -7, 10, -13
10. 1, 2, 4, 8, 16
12. $25, \frac{25}{2}, \frac{25}{3}, \frac{25}{4}, 5$
14. 1, 1, 3, 7 17
16. 0, 1, 4, 9, 16
18. $\frac{1}{2}, 1, \frac{5}{4}, \frac{7}{5}, \frac{9}{6}$
20. $\frac{1}{2}, \frac{1}{3}, \frac{1}{4}, \frac{1}{5}, \frac{1}{6}$
22. $a_n = 2n + 3$ or $a_{n+1} = a_n + 2$, where $a_1 = 5$
24. $a_n = 1/n$ or $a_{n+1} = 1/(a_n + 1)$, where $a_n = 1$

26. $a_n = 1 - (1/10)^n$

28. $a_{n+2} = a_n + a_{n+1}$, where $a_1 = \frac{1}{2}$ and $a_2 = \frac{1}{2}$

30. 2, 5, 4.1, 4.0, 4.0, 4.0, 4.0, 4.0, 4.0

32. $a_{n+1} = \dfrac{30/a_n + a_n}{2}$, $a_1 = 1$

8.2 PROGRESSIONS

Arithmetic Progressions

An *arithmetic progression* is a sequence of numbers in which the difference between any two consecutive elements is a fixed number. The fixed number is called the *common difference.*

In the arithmetic progression, 3, 5, 7, . . . , 19, the difference between any two consecutive elements is 2(5 − 3 = 2, 7 − 5 = 2).

In general notation, the elements of an arithmetic progression are:

1st term	a_1
2nd term	$a_1 + d$
3rd term	$a_2 + d = (a_1 + d) + d = a_1 + 2d$
4th term	$a_3 + d = (a_1 + 2d) + d = a_1 + 3d$
⋮	
*n*th term	$a_{n-1} + d = a_1 + (n - 1)\,d$

The general notation $a_n = a_1 + (n - 1)\,d$ can be used to generate any term of an arithmetic progression. As an example, suppose we want the 55th term of the sequence 3, 5, 7 . . . . This sequence has a common difference of 2 and the first term is 3. Substituting $n = 55$, $d = 2$, and $a_1 = 3$ into the general notation gives $a_{55} = 3 + (55 - 1)(2) = 3 + (54)(2) = 111$.

Example

Given an arithmetic progression whose fifth term is 37 and seventeenth term is 73. Find the first term and the common difference.

$$37 = a_1 + 4d \quad \text{and} \quad 73 = a_1 + 16d$$

This produces two linear equations in two unknowns, which can be solved by determinants or matrix algebra

$$a_1 = \frac{\begin{vmatrix} 37 & 4 \\ 73 & 16 \end{vmatrix}}{\begin{vmatrix} 1 & 4 \\ 1 & 16 \end{vmatrix}} \qquad d = \frac{\begin{vmatrix} 1 & 37 \\ 1 & 73 \end{vmatrix}}{\begin{vmatrix} 1 & 4 \\ 1 & 16 \end{vmatrix}}$$

or

$$\frac{1}{12}\begin{pmatrix} 16 & -4 \\ -1 & 1 \end{pmatrix}\begin{pmatrix} 1 & 4 \\ 1 & 16 \end{pmatrix}\begin{pmatrix} a_1 \\ d \end{pmatrix} = \frac{1}{12}\begin{pmatrix} 16 & -4 \\ -1 & 1 \end{pmatrix}\begin{pmatrix} 37 \\ 73 \end{pmatrix}$$

The solution is $a_1 = 25$ and $d = 3$.

A sequence with a common difference produces a linear relationship. The rate of change is constant. Many applications of this type of relationship exist.

Example

A library leases a photocopying machine for a flat fee of \$100 per month plus \$5.00 per 100 copies made. Express this as a sequence, each term representing the cost per hundred copies made.

$$\$100, \$105, \$110, \ldots$$

What payment is due if 1200 copies are made? The cost of 1200 copies is a_{13}:

$$a_{13} = 100 + (12)(5) = \$160$$

Geometric Progression

A *geometric progression* is a sequence of numbers in which the ratio between any two successive elements is a fixed number. A ratio suggests a fraction, and the fraction a_n/a_{n-1} must be fixed throughout the sequence. This ratio is referred to as r.

In general notation, a geometric progression, element by element is:

1st term	a_1
2nd term	$a_2 = a_1 r$
3rd term	$a_3 = a_2 r = (a_1 r)\, r = a_1 r^2$
4th term	$a_4 = a_3 r = (a_1 r^2)\, r = a_1 r^3$
⋮	
nth term	$a_n = a_{n-1} r = (a_1 r^{n-2})\, r = a_1 r^{n-1}$

The general notation $a_n = a_1 r^{n-1}$ can be used to generate any term of a geometric progression. As an example, suppose we want the 55th term of the sequence 5, 10, 20 . . . . This sequence has a common ratio of 2, and the first term is 5. Substituting the information into the general notation for $n = 55$,

$$a_{55} = (5)(2)^{54} = 6.0072 \times 10^{16}$$

—a very large number!

It is also possible to find the general term, first term, or ratio given any two terms of a geometric progression.

Example

Given $a_3 = \frac{4}{3}$ and $a_7 = \frac{64}{243}$. Use the general notation twice.

$$\frac{4}{3} = a_1 r^2 \quad \text{and} \quad \frac{64}{243} = a_1 r^6$$

Divide the second expression by the first to give the equation $r^4 = \frac{16}{81}$, or $r = \pm\frac{2}{3}$. Substituting this value into either equation gives $a_1 = 3$. There are two different possible answers: $3, -2, +\frac{4}{3}, -\frac{8}{9}, \ldots$ and $3, 2, \frac{4}{3}, \frac{8}{9}, \ldots$.

A sequence with a common ratio produces an exponential relationship. Many things in the world today behave in this manner—radioactivity, population, investment growth, to name a few.

Example

A grandparent invests \$1000 in an annuity for a newborn; the guaranteed annual rate of interest is 10%. What is the annuity worth in 18 years?

The sequence is \$1000, \$1100, \$1210, ... Thus $a_1 = \$1000$ (original investment) and $r = 1.10$ and $a_{19} = (1000)(1.10)^{18} = (\$1000)(5.56) = \$5560$.

EXERCISES

Determine if the following are arithmetic, geometric, or neither; determine the general term; and calculate the indicated term.

1. $4, 8, 16, \ldots; a_6$
2. $5, 7, 9, \ldots; a_9$
3. $7, 4, 1, \ldots; a_{10}$
4. $8, 4, 2, \ldots; a_8$
5. $\frac{1}{2}, \frac{3}{2}, \frac{5}{2}, \ldots; a_{16}$
6. $.3, .03, .003, \ldots; a_7$
7. $2, -2, 2, -2, \ldots; a_{47}$
8. $-5, 0, 5, \ldots; a_{10}$
9. $\frac{1}{2}, \frac{7}{16}, \frac{3}{8}, \frac{5}{16}, \ldots; a_{11}$
10. $1, \frac{1}{2}, \frac{1}{4}, \ldots; a_8$
11. $1, \frac{1}{2}, \frac{1}{3}, \ldots; a_{14}$
12. $.9, .99, .999, \ldots; a_8$
13. $1, 1, 2, 3, 5, \ldots; a_8$
14. $110, 121, 132, 143, \ldots; a_7$

Solve the following, using the notation of sequences.

15. A prospective employer offers a starting salary of \$12,000 with the choice of an increase of \$1500 a year or an increase of 10% of the employee's annual salary for a

satisfactory performance rating. Which option would you select? What is the better deal after 5 years? After 10 years?

16. The second and fifth terms of an arithmetic progression are $\frac{1}{5}$ and $\frac{16}{5}$. Find a_1 and d.
17. The second and fifth terms of a geometric progression are $\frac{1}{5}$ and $\frac{8}{5}$. Find a_1 and r.
18. $1000 is invested in a 12% annuity. Find the sequence representing the yearly value and determine the value after 10 years.
19. Depreciation for a vehicle costing $10,000 is 20% of the value of the vehicle at the beginning of that year. At what value is the vehicle being carried on the books after 5 years?
20. A basketball is dropped from a height of 100 feet, and each time it strikes the floor it rebounds 60% of the height from which it falls. How far does it rebound after the sixth bounce?
21. In a grain-storage elevator, shrinkage loss due to the drying process is 1% per week. A grower stores 3000 bushels of grain. How much is available for sale after 10 weeks?
22. A parking garage charges $1.00 plus 50¢ per hour after 1 hour. Write the sequence indicating charges and determine the amount due after $8\frac{1}{2}$ hours.
23. A grandfather clock strikes the hour and once on the half hour. Write the sequence indicating the number of strikes, assuming the strikes are recorded from 7 A.M. Is this an arithmetic or geometric sequence?
24. The salary schedule for a state agency begins at $10,400 per year with a $400 raise every 6 months on the job. How much would the salary of an employee be after 5 years? How long would it take to double the beginning salary?

Answers

2. Arithmetic; $a_n = 2n + 3; a_9 = 21$
4. Geometric; $a_n = 8(\frac{1}{2})^{n-1} = (\frac{1}{2})^{n-4}; a_8 = \frac{1}{16}$
6. Geometric; $a_n = .3(\frac{1}{10})^{n-1}; a_7 = .0000003$
8. Arithmetic; $a_n = 5n - 10; a_{10} = 40$
10. Geometric; $a_n = (\frac{1}{2})^{n-1}; a_8 = \frac{1}{128}$
12. Neither; $a_n = 1 - (\frac{1}{10})^n; a_8 = .99999999$
14. Arithmetic; $a_n = 11n + 99; a_7 = 176$
16. $d = 1, a = -\frac{4}{5}$
18. Geometric; $3,105.85
20. $a_n = 100(.6)^{n-1}; a_7 = 4.66$ feet
22. $5.00
24. After 5 years, $14,400; $13\frac{1}{2}$ years to double

8.3 SUMS OF SEQUENCES: SERIES

If $a_1, a_2, a_3, \ldots, a_n$ are the elements of a sequence, then the sum of these elements is called a *series*.

The sum $a_1 + a_2 + a_3 + \cdots + a_n$ is a series that can be symbolized by

$$\sum_{n=1}^{n} a_n$$

The task of finding a sum such as $1 + 2 + 3 + \cdots + 100$ would be dull and time-consuming; the answer would be prone to computational error. The sum $1 + 2 + 3 + 4 + \cdots 97 + 98 + 99 + 100$ is an arithmetic series with $a_1 = 1$, $a_{100} = 100$, and $d = 1$. If this series were ordered differently, the first and last terms grouped, the second and next to last terms grouped, and so on, the series would be: $(1 + 100) + (2 + 99) + (3 + 98) + (4 + 97) + \cdots + (50 + 51)$. Notice that each group totals 101 and there are 50 groups; therefore, the sum is $(50)(101) = 5050$.

To generalize this summation for any arithmetic sequence: $S_n = (a_1 + a_n) + (a_2 + a_{n-1}) + \cdots$ for $n/2$ terms, so $S_n = n/2(a_1 + a_n)$. That is, add the first and last terms of an arithmetic series and multiply the sum by half the number of terms. If the last term is not given, the formula $a_n = a_1 + (n - 1)d$ can be used to replace a_n. The formula for the sum in terms of a_1, d, and n is $a_1 n + [n(n - 1)/2]\, d$.

Example

Find the sum of the first 100 odd numbers.

$$S_{100} = (1)(100) + \frac{(100)(99)}{2}(2) = 10{,}000$$

Using summation notation, the example could also be written: Find

$$\sum_{n=1}^{100} (2n - 1).$$

A geometric series is the sum of the elements of a geometric progression. $S_n = a_1 + a_1 r + a_1 r^2 + a_1 r^3 + \cdots + a_1 r^{n-1}$. To determine a formula for this sum without the tedious task of adding element by element, let us multiply both sides of the sum by r.

$$rS_n = a_1 r + a_1 r^2 + a_1 r^3 + \cdots + a_1 r^{n-1} + a_1 r^n$$

From this, subtract $S_n = a_1 + a_1 r + a_1 r^2 + \cdots + a_1 r^{n-1}$.

$$rS_n - S_n = (a_1 r + a_1 r^2 + a_1 r^3 + \cdots + a_1 r^n) - (a_1 + a_1 r + a_1 r^2 + a_1 r^{n-1})$$

$$(r - 1)\, S_n = -a_1 + a_1 r^n$$

$$S_n = \frac{a_1 (r^n - 1)}{(r - 1)} \qquad r \neq 1$$

Example

Find the sum of the first ten terms of 2, 4, 8, 16 . . . .

This is a geometric series with $a_1 = 2$ and $r = 2$. Substituting into the formula,

$$S_{10} = \frac{2(2^{10} - 1)}{2 - 1} = \frac{2(1024 - 1)}{1} = 2046$$

This same problem can be written in summation notation as

$$\sum_{n=1}^{10} 2(2)^{n-1} = \sum_{n=1}^{10} 2^n$$

There are many other series that are useful and of interest to mathematicians. Values of the trigonometric functions, logarithms, and pi can all be determined by series. If a series cannot be classified as arithmetic or geometric, the individual elements must be added.

In chemistry, the half-life of a radioactive element is the time required for half of the given material to decay. For an element to have a half-life of 10 years, half of the amount present will be decayed in 10 years, half of the amount remaining will decay in the next 10 years, and so on. The element will never be completely gone!

If there is 1 pound of radioactive material, in 10 years there will be $\frac{1}{2}$ pound, in 20 years $\frac{1}{4}$ pound, and so on, giving 1, $\frac{1}{2}$, $\frac{1}{4}$, ...

How much material will there be after 50 years ($n = 5$ ten-year periods)?

$$a_1 = 1, \qquad r = \tfrac{1}{2} \qquad a_6 = (1)(\tfrac{1}{2})^5 = \tfrac{1}{32} \text{ pound remaining}$$

If the question were how much has decayed after 50 years, the sequence of decay would be $\frac{1}{2}$, $\frac{1}{4}$, . . . for five terms. The total amount decayed is:

$$\sum_{n=1}^{5} \left(\frac{1}{2}\right)^n = S_5 = \frac{\frac{1}{2}(\frac{1}{2}^5 - 1)}{\frac{1}{2} - 1} = \frac{\frac{1}{2}(\frac{1}{32} - 1)}{-\frac{1}{2}} = \frac{\frac{1}{2}(-\frac{31}{32})}{-\frac{1}{2}} = \frac{31}{32} \text{ pounds decayed}$$

EXERCISES

Calculate the sum of each of the following.

1. 4 + 8 + 16 + · · · ; 6 terms
2. 5 + 7 + 9 + · · · + 61
3. 7 + 4 + 1 + · · · ; 22 terms
4. 8 + 4 + 2 + · · · ; 8 terms
5. $\frac{1}{2} + \frac{3}{2} + \frac{5}{2} + \cdots$; 17 terms

6. $.3 + .03 + .003 + \cdots$; 7 terms
7. $2 + (-2) + 2 + \cdots$; 1579 terms
8. $-5 + 0 + 5 + \cdots 175$
9. $\frac{1}{2} + \frac{7}{16} + \frac{3}{8} + \cdots$; 7 terms
10. $1 + \frac{1}{2} + \frac{1}{4} + \cdots$; 20 terms
11. $.9 + .99 + .999 + \cdots$; 6 terms
12. $1 + \frac{1}{2} + \frac{1}{3} + \cdots$; 6 terms
13. $110 + 121 + 132 + \cdots$; 10 terms
14. $27 - 9 + 3 - 1 + \cdots$; 10 terms
15. A piece of machinery with initial value of $10,000 was depreciated yearly for 10 years. The depreciations were $1000, $900, $810, $719, ... How much of the original price had been depreciated after 10 years?
16. $1000 was invested at 12% compounded annually, so the interest recorded was $120, $134.40, $150.53, ... Find the total interest earned by the $1000 after 7 years.
17. A grandfather clock strikes the hour and once on the half hour. How many strikes are there in a 12-hour period?
18. A person earns $12,000 the first year of employment, and the salary increases $1000 each year over the next 10 years. Find the total salary earned in the 11 years.
19. A basketball is dropped from a height of 100 feet, and each time it strikes the floor it rebounds 60% of the height from which it falls. Find the total distance the ball has traveled when it strikes the floor the fifth time.
20. Mr. Jones borrows $1200.00 and agrees to repay it in 12 installments by paying $100.00 a month plus the interest at 1% per month on the unpaid balance. What is the total interest that he will pay?

Answers

2. 957
4. 15.9375
6. .3333333
8. 3145
10. 1.9999981
12. 2.45
14. 20.249657
16. $1210.68
18. 187,000
20. $78

8.4 BUSINESS APPLICATION UTILIZING SEQUENCE AND SERIES

Financial obligations are often met by installment payments. The partial payments must total to the obligation, interest, and carrying charges. It is often difficult to determine the rate of interest involved in these transactions because the obligation changes monthly.

Example

A computer sells for \$615.00. It can be purchased for \$75.00 down and 12 payments of \$50.00 each. Find the rate of interest charged.

\$615 − \$75 = \$540	Balance of the obligation
\$50 × 12 = \$600	Total installment payments
\$600 − \$540 = \$60	Interest and carrying charges

Add interest and carrying charges to the balance of the obligation: \$60 + \$540 = \$600 total balance owed, month 1. The balances are \$600, \$550, \$500, . . . , \$50. The sum of the balances for each month is the same as the amount owed for 1 month.

$$S_{12} = \frac{12}{2}(600 + 50) = 3900$$

Use the formula $I = prt$; solve for r:

$$r = \frac{I}{pt} = \frac{60}{(3900)(\frac{1}{12})} = 18.5\% \text{ annual percentage rate}$$

If interest has not been added into the original transaction but is added each month, the rate of interest is given by the *constant ratio formula.*

$$\text{Rate} = \frac{2(\text{number of payments in one year})(\text{amount of interest})}{(\text{unpaid balance})(\text{number of payments} + 1)}$$

In the last example, if the unpaid balance was \$540 in month 1, the rate of interest is calculated by:

$$\text{Rate} = \frac{2(12)(60)}{(540)(13)} = 20.5\%$$

The rate is slightly higher because the obligation is lower. It does not include the interest. There are other formulas used that yield different rates of interest; the rate depends on the formula used by the company calculating the loan.

Many financial investments or obligations are made for long periods of time, 5 years or more, and simple interest is not applicable. Compound interest is interest that accrues on both the principal and the interest in the account. The compound amount increases throughout the term. The difference between the compound amount and the original principal is the compound interest.

Example

Find the compound amount and the compound interest on \$100,000 invested for 5 years at 10% compounded annually.

Interest earned in year 1 is \$10,000. The compound amount is \$110,000.
Interest earned in year 2 is \$11,000. The compound amount is \$121,000.
Interest earned in year 3 is \$12,100. The compound amount is \$133,100.
Interest earned in year 4 is \$13,310. The compound amount is \$146,410.
Interest earned in year 5 is \$14,641. The compound amount is \$161,051.

The compound interest is \$161,051 − 100,000 = \$61,051. If we let P represent the amount invested:

The compound amount after 1 year is $P(1 + i)$.
The compound amount after 2 years is $P(1 + i)(1 + i) = P(1 + i)^2$.

The compound amount forms a geometric progression whose nth term is $A = P(1 + i)^n$, where i represents the interest rate for the compound period and n is the number of compounding periods.

In this example, P = \$100,000, i = 10%, and $n = 5$.

$$\text{Compound amount} = 100{,}000(1 + .1)^5 = 100{,}000(1.1)^5 = \$161{,}051$$

Suppose this example is restated: Find the compound amount and the compound interest on \$100,000 invested for 5 years at 10% compounded quarterly. Then, $P = 100{,}000$, $i = .10/4 = .025$, and $n = 5 \times 4 = 20$. Quarterly compounding indicates that the interest rate be divided by 4 and the years be multiplied by 4.

$$A = 100{,}000(1 + .025)^{20} = 100{,}000(1.025)^{20} = 100{,}000(1.6386165)$$
$$= \$163{,}861.65$$

This is \$2810.65 more than the amount in the example.

The formula $A = P(1 + i)^n$ can be solved for P.

$$P = \frac{A}{(1 + i)^n}$$

This is the formula for *present value*. Another form of this formula is

$$P = A(1 + i)^{-n}$$

Example

Suppose you are the proud parents of a child who promises to be the brightest about! You have 18 years to save for the college education that your bright child deserves; with inflation and costs skyrocketing, you want to have \$50,000 for your child in 18 years. How much must you invest today at 12% compounded monthly to ensure \$50,000 in 18 years? In other words, what is the present value of the \$50,000? A = 50,000, the compound amount; $i = .12/12 = .01$; and $n = 18 \times 12 = 216$.

$$P = \frac{50{,}000}{(1 + .01)^{216}} = \frac{50{,}000}{8.5786} = \$5{,}828.45$$

EXERCISES

1. A washing machine is purchased for \$49.95 down and \$37.50 a month for 12 months. The sale price of the machine was \$449.95. What rate of interest is being charged if the interest and carrying charges are added to the obligation?
2. A loan company advertises a loan of \$1000 to be repaid in 24 monthly payments of \$55. Interest and carrying charges are added to the amount of the loan. What rate of interest is being charged?
3. An automobile was advertised at a price of \$1500—\$300 down and \$60 a month for 24 months. Find the rate of interest by the constant ratio formula.
4. A diamond ring lists for \$780.00 and can be purchased for \$180 down and \$60 a month for 12 months. Find the rate of interest charged using the constant ratio formula.
5. Find the compound amount and compound interest on \$5000 at 6% interest, compounded yearly for 5 years.
6. Find the compound amount and compound interest on \$2000 at 8% interest, compounded semiannually for 4 years.
7. Find the compound amount and compound interest on \$3000 at 8% interest, compounded quarterly for 3 years.
8. Find the compound amount and compound interest on \$5000 at 9% interest, compounded monthly for 2 years.
9. Find the present value of \$10,000, compounded quarterly at 8% interest, for 5 years.
10. Find the present value of \$12,000, compounded semiannually at 10% interest, for 6 years.

Answers

2. 23.3%
4. 36.9%
6. \$2737.14 compound amount; \$737.14 compound interest
8. \$5982.07 compound amount; \$982.07 compound interest
10. \$6682.05 present value

CHAPTER 8 REVIEW EXERCISES

Write the general notation for Examples 1-20. If the sequence is arithmetic, find the sum of the first 12 terms. If the sequence is geometric, find the sum of the first 7 terms. If the sequence is neither, write the sum of the first 10 terms in sigma notation.

1. 49, 46, 43, . . .
2. 36, 18, 9, . . .
3. \$100, \$104, \$108.16, . . .
4. 2000, 2040, 2080, . . .

5. 12, 13, 14, . . .
6. 6, 2, $\frac{2}{3}$, . . .
7. 5, 10, 15, 20, . . .
8. 5, 5, 10, 15, . . .
9. $\frac{1}{2}, \frac{1}{4}, \frac{1}{6}$, . . .
10. .8, .96, .992, . . .

Write the first five terms of the following sequences. If the sequence is arithmetic, circle d in the general notation. If the sequence is geometric, circle r in the general notation.

11. $a_n = 5n + 1$
12. $a_n = 6 - 2n$
13. $a_n = (-1)^{n-1}(2)^n$
14. $a_n = (-1)^{2n}(3)^{-n}$
15. $a_n = \dfrac{n}{n+1}$
16. $a_n = (2n - 1)^2$
17. $a_{n+1} = 2a_n + 1, a_1 = -1$
18. $a_{n+2} = \dfrac{a_{n+1}}{a_n}, a_1 = 5, a_2 = \dfrac{5}{7}$
19. A superball is dropped from 30 feet and rebounds half the height fallen each time it bounces. How far has the superball traveled after five bounces?
20. Find the rate of interest charged for an installment loan on \$499.95 with \$49.95 down and \$20 a month for 2 years. The interest and carrying charges are added in at the time of the purchase.

Chapter 8 Review Answers

2. $a_n = 36(\frac{1}{2})^{n-1}; S_7 = 71.4375$
4. $a_n = 40n + 1960; S_{12} = 26.640$
6. $a_n = 6(\frac{1}{3})^{n-1}; S_7 = 8.996$
8. $a_{n+2} = a_n + a_{n+1}; a_1 = 5; a_2 = 5; \sum_{n=1}^{8} (a_n + a_{n+1}) + a_1 + a_2$
10. $a_n = 1 - (.2)^n; \sum_{n=1}^{10} (1 - .2)^n$
12. $4, 2, 0, -2, -4; a_n = 6 \enclose{circle}{-2}\, n$
14. $\frac{1}{3}, \frac{1}{9}, \frac{1}{27}, \frac{1}{81}, \frac{1}{243}; a_n = (-1)^{2n} \enclose{circle}{\left(\frac{1}{3}\right)}^{n}$
16. 1, 9, 25, 49, 81
18. $5, \frac{5}{7}, \frac{1}{7}, \frac{1}{5}, \frac{7}{5}$
20. 12%

Linear Programming

CHAPTER 9

9.0 INTRODUCTION

Linear Programming is a relatively new development in applied mathematics that came into being during World War II. An American mathematician, George Dantzig, is credited with the theory of linear programming.

Linear programming is the mathematics used to solve large-scale problems concerned with optimal values for profits, efficiency, and productivity, subject to restrictions or constraints such as availability of labor and materials. The restrictions are linear functions; the values of the restrictions are positive numbers. The problem is similar to solving a large linear system that involves many possible solutions. The optimal solution is required.

In order to give some insight into the nature of linear programming, a graphical solution technique is presented first. The nature of graphing dictates that only two or three variables can be used. The simplex method of solution is the method involving an algebraic approach. It works for any number of variables and is the method employed in computer solutions.

This chapter is designed to:

1. Provide the graphical solution for two-variable problems to give an understanding of linear programming and its use.
2. Provide the simplex method of solution for maximization to give an algebraic approach to a linear-programming problem and suggest the computer use in solutions.

3. Provide the duality theorem to give a method of solution for minimization problems.

9.1 GRAPHICAL SOLUTION FOR LINEAR PROGRAMMING

To illustrate the notion of a problem that utilizes linear programming, let us consider the following example.

The ABC Manufacturing Company produces products A and B. Three machines are used in the manufacture of these products.

Product A requires 3 hours on machine I, 7 hours on machine II and 10 hours on machine III.

Product B requires 4 hours on machine I, 5 hours on machine II and 4 hours on machine III.

Machine I has at most 120 production hours per week.

Machine II has at most 175 production hours per week.

Machine III has at most 200 production hours per week.

Business is good and any number of products A and B can be sold. Profit from product A is $5 per item; profit is $3 for Product B. Find the production combination to maximize profit. All the above information can be put into a table.

	A	B	Maximum time available
Machine I	3	4	120
Machine II	7	5	175
Machine III	10	4	200
Profit	5	3	

The values for A and B represent the number of items produced. The algebraic relationships from the table are:

$$3A + 4B \leq 120$$

$$7A + 5B \leq 175$$

$$10A + 4B \leq 200$$

$$5A + 3B \qquad \text{Profit}$$

The inequality is used because at most indicates less than or equal to; only greater than is unacceptable.

Common sense indicates that as many A as possible should be produced because of the higher profit for the product. However machine III can produce only 20 A and no B because of the high demand of Product A on machine III time.

For the graphical solution, xy-coordinate axes are drawn, and because linear programming is applicable only for positive solutions, only the upper-right quadrant is needed. Either axis can be labeled A and B; however, they must be labeled. To graph $3A + 4B \leq 120$, first draw the graph of $3A + 4B = 120$. Use the intercept method to draw the straight line representing $3A + 4B = 120$. Let $A = 0$ and solve for B which is 30. Let $B = 0$ and solve for A which is 40. Locate these points on the two axes to determine the line. This line divides the graph into three distinct areas. The area above the line is

$$3A + 4B > 120$$

The area on the line is $3A + 4B = 120$. The area below the line is $3A + 4B < 120$. Because the inequality being graphed has both equal and less than symbols, the area represented by the inequality is the line and everything below it. (If the line is not to be included, a dotted line is used to denote the area up to but not including the line.) The area is indicated by /// lines. On the same axes, graph the area represented by $7A + 5B \leq 175$. The intercepts are $B = 35$ (when $A = 0$) and $A = 25$ (when $B = 0$). Locate these points on the axis, draw a solid line (because of the equality) and shade the area below the line with \\\ lines. On the same axes, graph the area represented by $10A + 4B \leq 200$. The intercepts are $A = 20$ (when $B = 0$) and $B = 50$ (when $A = 0$). Locate these points on the axis, draw a solid line, and shade the area below it.

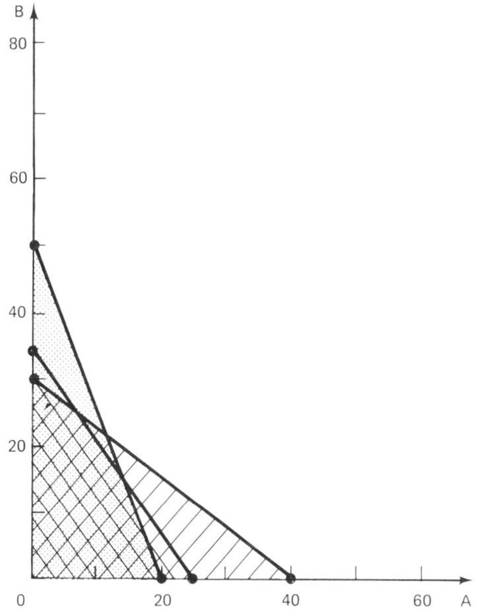

The area containing all three types of shading is the area satisfying the three constraints. This area contains an infinite number of points, all of which satisfy the conditions when substituted in the constraints.

A cleaned-up version of the graph helps in visualizing the graphical solution. The shaded area forms a polygon with five vertices. Some of these values have to be approximated, but all can be named. For clarity, the vertices have been identified as X_1, X_2, X_3, X_4, and X_5

X_1: $A = 0$ and $B = 0$

X_2: $A = 20$ and $B = 0$

X_3: $A = 15$ and $B = 15$

X_4: $A = 6$ and $B = 26$

X_5: $A = 0$ and $B = 30$

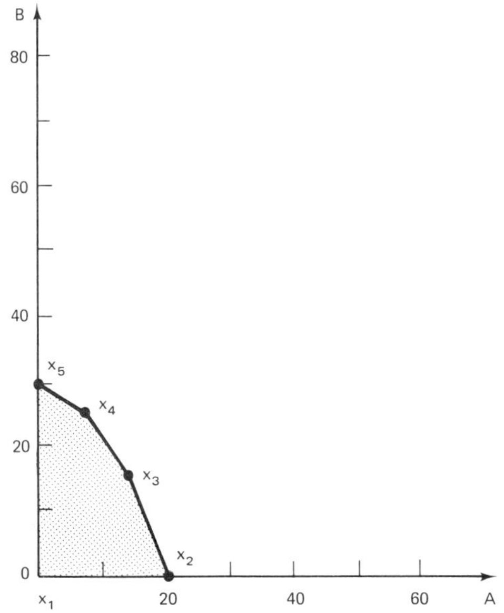

If maximum and minimum values of the function exist, they must occur at vertices.

The theorem stating this fact is cited without proof: Every linear function defined over the set of points of a convex polygon reaches its maximum value at one vertex of the polygon and its minimum value at another vertex of the polygon.

The function of the example being graphed is to maximize the profit.

The profit is dependent on A and B: $5A + 3B$. This function is evaluated at each of the vertices of the polygon (X_1, X_2, X_3, X_4, X_5).

X_1:	$A = 0$,	$B = 0$	$5A + 3B = 5(0) + 3(0) = 0$	Minimum
X_2:	$A = 20$,	$B = 0$	$5A + 3B = 5(20) + 3(0) = 100$	
X_3:	$A = 15$,	$B = 15$	$5A + 3B = 5(15) + 3(15) = 120$	Maximum
X_4:	$A = 6$,	$B = 26$	$5A + 3B = 5(6) + 3(26) = 108$	
X_5:	$A = 0$,	$B = 30$	$5A + 3B = 5(0) + 3(30) = 90$	

The numbers indicate the greatest profit is realized when 15 of A and 15 of B are manufactured. The least profit is realized when none of either is manufactured.

Summary

1. Write the constraints as linear inequalities and the objective function as an algebraic expression.
2. Graph each of the constraints, shading the areas representing the solution sets. The intersection of the shaded areas is the solution area.
3. Identify the vertices of the solution area.
4. Substitute the values at the vertices into the objective function. The vertex giving the largest value is the required solution.

EXERCISES

Graph each of the following, identify the values associated with each vertex, and find which values maximize and minimize the objective function.

1. $x + y \leq 8$
   $5x + y \leq 15$
   $2x + y \leq 9$
   Objective function $3x + y$
2. $x + 3y \leq 12$
   $2x + y \leq 6$
   $x \leq 2$
   Objective function $x + 7y$
3. $x > 1$
   $2x + y \leq 7$
   $2x + 3y \leq 9$
   Objective function $3x + 4y$

4. $7x + 6y \leq 17$
   $x + 4y \leq 8$
   $9x + 4y \leq 18$
   Objective function $5x + 4y$
5. $6x + 7y \leq 21$
   $2x + 5y \leq 12$
   $3x + 2y \leq 9$
   Objective function $5x + 4y$
6. $x + 3y \leq 340$
   $3x + 4y \leq 600$
   $x + y \leq 180$
   Objective function $7x + 7y$
7. A manufacturer of televisions and radios must produce at most 350 televisions and 700 radios each week. Twenty-four worker-hours are required to assemble a television. Five worker-hours are required to assemble a radio. The company has at most 10,000 worker-hours available each week. The profit is \$25.00 from each television set and \$10.00 from each radio. Find the production of televisions and radios that will maximize profit.
8. A wood craftsperson makes tables and beds. Each table requires 3 hours on the lathe and 5 hours on the joiner. Each bed requires 14 hours on the lathe and 3 hours on the joiner. There are 75 hours available on the lathe each week and 60 hours on the joiner. The product is unique and all products produced can be sold with a profit of \$50 on each table and \$90 on each bed. Determine the production output to maximize the profit.
9. Two plants manufacture products A and B for the same firm. Plant 1 can make product A in 10 hours and product B in 8 hours. Plant 2 can make product A in 7 hours and product B in 9 hours. The maximum hours available in plant 1 is 1200 hours; in plant 2 there are 1000 hours available. The profit from product A is \$15 and from product B is \$16. How would you schedule production in the two plants to maximize profit?
10. A man has \$100,000 to invest in stocks and bonds. Stocks produce an average return of 10%; the average yield of bonds is 12%. The stockbroker advises the man to invest not more than \$90,000, leaving \$10,000 for a savings account. The stockbroker also advises at least twice as much in the stock markets as in bonds to help fight inflation. Find the most profitable investment mix.

Answers

2.

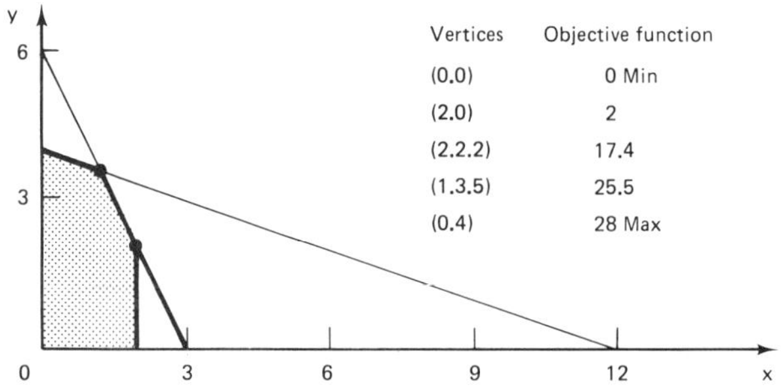

Vertices	Objective function
(0.0)	0 Min
(2.0)	2
(2.2.2)	17.4
(1.3.5)	25.5
(0.4)	28 Max

4.

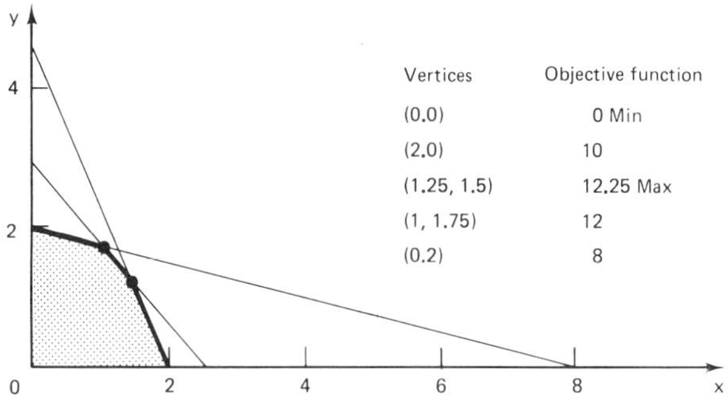

Vertices	Objective function
(0.0)	0 Min
(2.0)	10
(1.25, 1.5)	12.25 Max
(1, 1.75)	12
(0.2)	8

6.

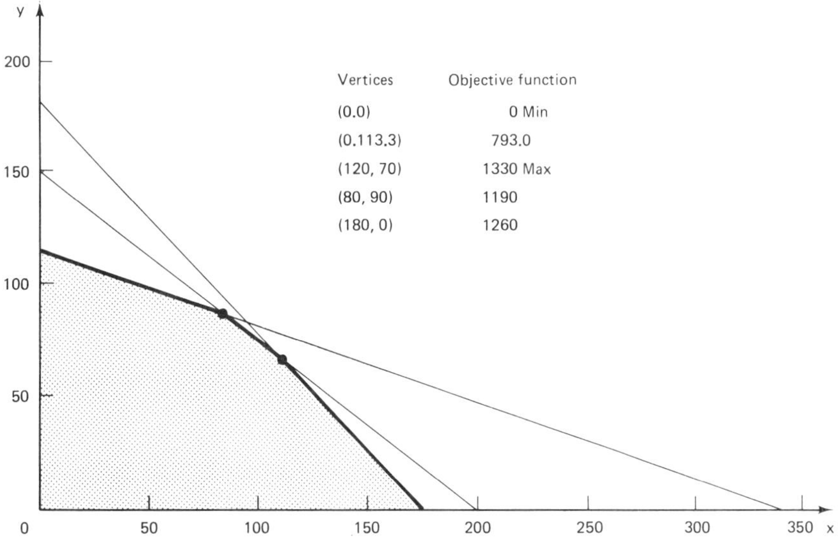

Vertices	Objective function
(0.0)	0 Min
(0.113.3)	793.0
(120, 70)	1330 Max
(80, 90)	1190
(180, 0)	1260

8. Inequalities: $3T + 14B \leq 75, \qquad 5T + 3B \leq 60$

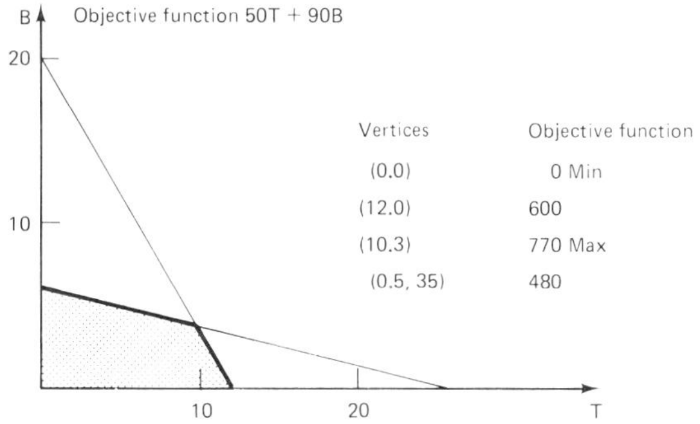

Vertices	Objective function
(0.0)	0 Min
(12.0)	600
(10.3)	770 Max
(0.5, 35)	480

10.

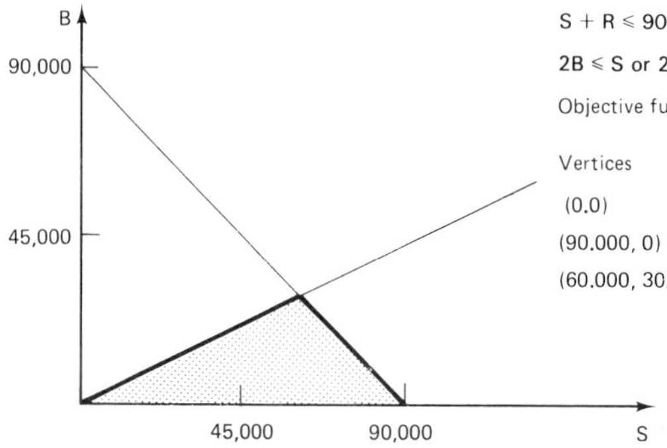

S + R ≤ 90.000

2B ≤ S or 2B − S ≤ 0

Objective function 0.12B + 0.10S

Vertices	Objective function
(0.0)	0 Min
(90.000, 0)	9000
(60.000, 30.000)	9600

9.2 THE SIMPLEX METHOD

The problem used to demonstrate the graphical solution is used to illustrate the *simplex method.* The original set of constraints were

$$3A + 4B \leq 120$$

$$7A + 5B \leq 175$$

$$10A + 4B \leq 200$$

Objective function $\quad 5A + 3B = \text{profit}$

Inequalities have many solutions; equations are more limited in the number of solutions. It is necessary to write each of the constraints as equations, and this is achieved by adding something—s_1, s_2, or s_3—to each constraint. These are called *slack variables* and always represent positive quantities. Thus

$$3A + 4B + s_1 = 120$$

$$7A + 5B + s_2 = 175$$

$$10A + 4B + s_3 = 200$$

This information will be placed into a table, called the *initial tableau.* It resembles an array in matrix algebra.

A	B	s_1	s_2	s_3	
3	4	1	0	0	120
7	5	0	1	0	175
10	4	0	0	1	200
-5	-3	0	0	0	0

The objective function is added as the last row. To make the simplex method work, the profit values must be negated.

The initial tableau reflects the condition that $A = 0$; $B = 0$; $s_1 = 120$, which is the amount of unused time by Machine 1; $s_2 = 175$, which is the amount of unused time of machine II; and $s_3 = 200$, the amount of unused time of machine III. The profit is 0. This represents the vertex (0, 0) in the graphical solution.

It is most sensible to start making the product with the greatest profit. Thus the choice of column A as the initial working area. Each entry in the rightmost column is divided by the A column entries. These ratios are compared and the smallest ratio determines the first pivot element.

$$\frac{120}{3} \text{ gives } 40,$$

$$\frac{175}{7} \text{ gives } 25, \text{ and}$$

$$\frac{200}{10} \text{ gives } 20.$$

20 is the smallest ratio; the first pivot element is 10.

	A	B	s_1	s_2	s_3	
	3	4	1	0	0	120
	7	5	0	1	0	175
Pivot element	(10)	4	0	0	1	200
	-5	-3	0	0	0	0

The goal is to replace the 10 with a 1 and all of the other elements in column A with 0. This is achieved by *elementary row operations.* Divide each element in row 3 by 10:

1 .4 0 0 .1 20

This is called the *working row* and is used to get 0 in every other location in the column.

To get a 0 in row 1 column 1, add row 1 to -3 times the working row.

3	4	1	0	0	120	
-3	-1.2	0	0	-.3	-60	
0	2.8	1	0	-.3	60	New row 1

To get 0 in row 2, column 1, add row 2 to -7 times the working row.

7	5	0	1	0	175	
-7	-2.8	0	0	-.7	-140	
0	2.2	0	1	-.7	35	New row 2

To get a new row 4 with 0 in column 1, multiply the working row by 5 and add it to row 4.

-5	-3	0	0	0	0	
5	2	0	0	.5	100	
0	-1	0	0	.5	100	New profit number

The new tableau produced by these elementary row operations is:

	A	B	s_1	s_2	s_3	
	0	2.8	1	0	-.3	60
	0	(2.2)	0	1	-.7	35
A	1	.4	0	0	.1	20
	0	-1	0	0	.5	100

Notice that the column heading became a row heading when the pivot occurred. This tableau states that when 20 of A and 0 of B are manufactured, the profit is $100; the unused portion of time for machine I is 60 hours and for machine II is 35 hours. This is the same as vertex X_2 in the graphical solution. Because there is still a negative value in the objective function row, this profit can be improved.

Column 2 is selected because of the -1. The ratios 60/2.8, 35/2.2, and 20/.4 are checked. The smallest number among 21.4, 15.9, and 50 is selected. The element involved with 15.9 is 2.2, so 2.2 is the new pivot point.

The first step is to make the new pivot point equal to 1; this is achieved by dividing each element of the second row by 2.2.

0	1	0	.45	-.32	15.9	New working row

To replace 2.8 by 0, multiply the working row by -2.8 and add it to row 1.

0	-2.8	0	-1.2	.9	-44.5	
0	2.8	1	0	-.3	60	
0	0	1	1.2	.6	15	New row 1

To replace .4 by 0, multiply the working row by -.4 and add it to row 3.

0	-.4	0	-.18	.13	-6.4	
1	.4	0	0	.1	20	
1	0	0	.18	.23	14	New row 3

To replace 1 by 0, add the working row and row 4.

0	1	0	.45	.32	16	
0	-1	0	0	.5	100	
0	0	0	.45	.82	116	New profit

The new tableau produced by these elementary row operations is:

	A	B	s_1	s_2	s_3	
	0	0	1	1.2	.6	15
B	0	1	0	.45	−.32	16
A	1	0	0	.18	.23	14
	0	0	0	.45	.82	116

This is the final tableau because all entries in the last row are positive. The maximum value of the profit is 116, the entry in the lower right-hand corner. The tableau also shows that the maximum value is attained with $A = 14$ and $B = 16$. The top row indicates that machine I has 15 hours of unused machine time.

The solution attained from the simplex method corresponds to vertex X_3 in the graphical solution. A slight variation occurs in values between the graphical and algebraic solutions. Graphing is always approximate because of the difficulty in reading exact values from a graph.

The simplex method involves only the operations of division, multiplication, and addition. Whenever these elementary operations are used, the computer becomes a great resource and timesaver. Without computers, linear programming would not have gained wide acceptance as a reasonable mathematical tool.

Summary

1. Write each linear constraint as an equation, introducing slack variables $s_1, s_2, \ldots, s_n$ for the n inequalities.
2. Set up the initial tableau based on the constraints written as equations. The objective function written with negative coefficients is in the bottom row.
3. Find the pivot element by choosing the column with the most negative element in the function row. Choose the row containing the element that produces the smallest ratio with the rightmost number in that row.
4. Perform elementary row operations on Tableau 1 until the pivot element becomes 1 and all other elements in the pivot column are zero.
5. Repeat Steps 3 and 4 until all elements in the bottom row are nonnegative.
6. The maximum value is the element in the last row and the right column. The values of the variables that produce the maximum value are the numbers in the right-hand column, associated with the variable that has a 1 in that row and column.

EXERCISES

1. Write the following constraints as equations, introducing slack variables.

$$2x_1 + 3x_2 + 15x_3 \leq 158$$
$$4x_1 + 6x_2 \leq 86$$
$$3x_1 + 10x_3 \leq 120$$

Maximize $15x_1 + 20x_2 + 17x_3$.

2. Write the following constraints as equations, introducing slack variables.

$$x_1 + 2x_2 + 5x_3 \leq 14$$
$$5x_1 + 7x_2 + x_3 \leq 22$$
$$x_3 \leq 2$$

Maximize $3x_1 + 2x_2 + x_1$.

3. Construct the initial tableau for Exercise 1 and select the pivot element.
4. Construct the initial tableau for Exercise 2 and select the pivot element.
5. Perform elementary operations on the tableau in Exercise 3 to make the pivot element 1 and all other elements in the column 0.
6. Perform elementary operations on the tableau in Exercise 4 to make the pivot element 1 and all other elements in the column 0.
7. Choose second pivot entry in Exercise 5 and perform the elementary operations to make this entry 1 and all other entries in the column 0.
8. Choose second pivot entry in Exercise 6. Notice there are two entries that produce the same ratio; you may use either as a pivot element. Perform the elementary operations to make this entry 1 and all other entries in the column 0.
9. Interpret the results of the second pivot in Exercise 7.
10. Interpret the results of the second pivot in Exercise 8.

11 -20. Repeat the exercises in Section 9.1 using the simplex method of solution.

Answers

2. $x_1 + 2x_2 + 5x_3 + s_1 = 14$

$$5x_1 + 7x_2 + x_3 + s_2 = 22$$
$$x_3 + s_3 = 2$$

4.

x_1	x_2	x_3	s_1	s_2	s_3	
1	2	5	1	0	0	14
(5)	7	1	0	1	0	22
0	0	1	0	0	1	2
-3	-2	-1	0	0	0	0

6.

x_1	x_2	x_3	s_1	s_2	s_3	
0	.6	4.8	1	−.2	0	9.6
1	1.4	.2	0	.2	0	4.4
0	0	1	0	0	1	0
0	2.2	−.4	0	.6	0	13.2

8. The second pivot element is 1 or 4.8. Choose 1 in row 3.

	x_1	x_2	x_3	s_1	s_2	s_3	
	0	.6	0	1	−.2	−4.8	0
x_1	1	1.4	0	0	.2	−.2	4
x_3	0	0	1	0	0	1	2
	0	2.2	0	0	.6	.4	14

10. The maximum value of this problem is 14 and this value is achieved when $x_1 = 4, x_3 = 2$, and $x_2 = 0$.

12. Final tableau

x	y	s_1	s_2	s_3	
$\frac{1}{3}$	1	$\frac{1}{3}$	0	0	4
$\frac{5}{3}$	0	$-\frac{1}{3}$	1	0	2
1	0	0	0	1	2
$\frac{4}{3}$	0	$\frac{7}{3}$	0	0	28

Solution: $x = 2, y = 4$
Maximum value: 28

14.

	x	y	s_1	s_2	s_3	
y	0	1	$\frac{9}{26}$	0	$-\frac{7}{26}$	1.04
	0	0	$-\frac{32}{26}$	1	$\frac{1}{2}$	2.3
x	1	0	$-\frac{2}{13}$	0	.22	1.5
	0	0	.73	0	.09	12

Solution: $x = 1.5, y = 1.04$
Maximum value: 12

16. Two equal solutions:

$x = 180, y = 0$; Maximum value of 1260

$x = 100, y = 80$; Maximum value of 1260

18.

	T	B	s_1	s_2	
B	0	1	.08	−.05	3.36
T	1	0	−.05	.23	10
	0	0	5	6.9	782.4

Because decimals have no meaning, 3 beds, 10 tables.
Maximum profit \$770.

20.

	S	B	s_1	s_2	
S	1	0	$\frac{2}{3}$	$-\frac{1}{3}$	60,000
B	0	1	$\frac{1}{3}$	$\frac{2}{3}$	30,000
	0	0	.10	-.02	9600

Invest \$60,000 in stocks, 30,000 in bonds. Income: \$9600

9.3 MINIMUM LINEAR PROGRAMMING PROBLEM

In certain cases, linear-programming techniques are used to find the minimum value of a function rather than the maximum. If the objective function relates to cost, it is desirable to minimize the cost. Graphical techniques for minimization problems are similiar to those used in Section 9.1. The feasible area for the constraints is usually an unbounded area. The minimization problem involves greater than and equal to inequalities in the constraints. The minimum amount is given; this can be exceeded.

Example

It is required that some tablets containing specified amounts of three vitamins be purchased. The druggist wishes to minimize the costs of these tablets. She has two possible products, Pep and Superpep. Pep contains 4 units of vitamin 1, 3 units of vitamin 2, and 4 units of vitamin 3; the cost is 6¢ a tablet. Superpep contains 5 units of vitamin 1, 6 units of vitamin 2, and 3 units of vitamin 3; the cost is 8¢ per tablet. The government has determined the minimum daily requirement of vitamin 1 is 30 units, of vitamin 2 is 29 units, and of vitamin 3 is 21 units. What is the minimum daily cost for these vitamins?

Make a table of information.

	Pep	Superpep	Min requirement
Vitamin 1	4	5	30
Vitamin 2	3	6	29
Vitamin 3	4	3	21
Cost	6	8	

Written in algebraic form, the information is:

$4P + 5S \geq 30$

$3P + 6S \geq 29$

$4P + 3S \geq 21$

Cost: $6P + 8S$

Intercepts: $S = 6$ when $P = 0$ and $P = 7.5$ when $S = 0$; $S = 4.8$ when $P = 0$ and $P = 9.7$ when $S = 0$; $S = 7$ when $P = 0$ and $P = 5.25$ when $S = 0$

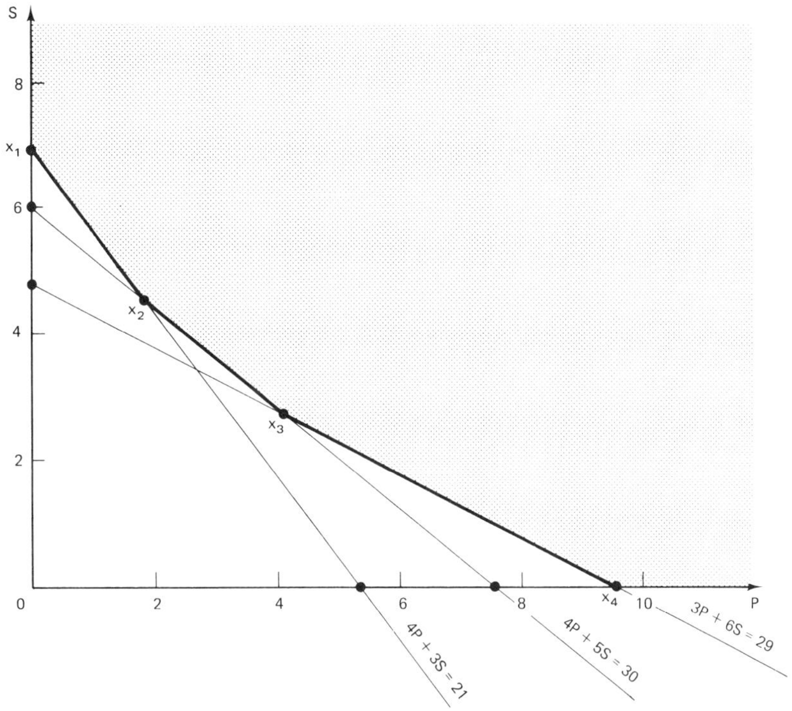

The graphical representation of the problem is shown. Notice the shaded area is not bounded above. This problem is concerned with minimum cost; therefore, the lower part of the shaded area is the area with which to be concerned. The four vertices are

$$X_1: \quad P = 0, \quad S = 7$$

$$X_2: \quad P = 2, \quad S = 4.5$$

$$X_3: \quad P = 4, \quad S = 3$$

$$X_4: \quad P = 9.7, S = 0$$

Because these are tablets, the decimals must be rounded up.

$P = 0, \quad S = 7$	$\text{Cost} = 6P + 8S = 56$	
$P = 2, \quad S = 5$	$\text{Cost} = 6(2) + 8(5) = 52$	
$P = 4, \quad S = 3$	$\text{Cost} = 6(4) + 8(3) = 48$	Minimum cost
$P = 10, S = 0$	$\text{Cost} = 6(10) + 8(0) = 60$	

Each maximum linear-programming problem has a corresponding minimum problem called its *dual*. Likewise, each minimum problem has a corresponding maximum problem, also called its *dual*. The dual is determined by a transpose of the matrix of the original problem. Our objective in converting a minimum problem to its dual maximum problem is to enable us to solve using the techniques already learned. The theorem that states that this is possible is called the *duality theorem*, and it is related without proof: The function f of a maximum linear programming problem reaches a maximum value if and only if its dual function reaches a minimum value.

The minimum problem in this section is:

$$4P + 5S \geq 30$$
$$3P + 6S \geq 29$$
$$4P + 3S \geq 21$$

The minimum objective: $6P + 8S$

This can be written in the format of the initial tableau before the slack variables are introduced as shown on the left. The transpose, or dual, is on the right.

4	5	30
3	6	29
4	3	21
6	8	0

4	3	4	6
5	6	3	8
30	29	21	0

The dual maximum problem, written with new variables W_1, W_2, and W_3, is:

$$4W_1 + 3W_2 + 4W_3 \leq 6$$
$$5W_1 + 6W_2 + 3W_3 \leq 8$$

Maximize $30W_1 + 29W_2 + 21W_3$. The initial tableau for the dual is:

	W_1	W_2	W_3	s_1	s_2		
	(4)	3	4	1	0	6	
	5	6	3	0	1	8	
	-30	-29	-21	0	0	0	The pivot point is column 1, row 1.
W_1	1	.75	1	.25	0	1.5	
	0	(2.25)	-2	-1.25	1	.5	
	0	-6.5	9	7.5	0	45	The pivot point is column 2, row 2.

W_1	1	0	1.66	.66	−.33	1.335
W_2	0	1	−.88	−.55	.44	.222
	0	0	3.23	3.92	2.86	46.43

Because the dual is being solved, the solution is in the bottom row rather than in the right-hand column; P is in the s_1-column and has a value of 3.92, S is in the s_2-column and has a value of 2.86, and the minimum cost is 46.43¢. To compare the solution of $S = 2.86$ and $P = 3.92$ with the graphical solution, whole unit answers are needed. The solution $S = 3$ and $P = 4$ would alter the cost figure of 46.43¢ slightly.

There are many other things that can be read from a final tableau and ways to affect change to the original constraints and still use the final tableau to predict solutions. If linear programming techniques can be utilized in the solution of problems, a complete exposition of the subject should be utilized as a reference.

EXERCISES

1. Given

$$3A + 4B \geq 120$$

$$5A + 2B \geq 100$$

$$7A + 5B \geq 210$$

The minimum function $3A + 2B$ is subject to the given constraints. Solve the system graphically.

2. Write the dual maximum problem for the Exercise 1, using W_1, W_2, and W_3 as variables in the dual.
3. Write the initial tableau of the dual problem and find the first pivot element.
4. Solve the dual problem, and—using the final tableau—give the values of A and B that result in the minimum value for $3A + 2B$.
5. A farmer discovers his crop needs at least 15 units of nitrogen and 8 units of phosphate per acre to be fertilized properly. Two products, Viga and Pro, are available. The content of the two products are as follows:

   Viga has 4 units of nitrogen per bag, 1 unit phosphate per bag, and costs $4 per bag.

   Pro has 3 units of nitrogen per bag, 3 units of phosphate per bag, and costs $6 per bag.

   Find the least-expensive mixture to satisfy the minimum requirements.

6. Accommodations must be provided for at least 500 athletes at the Olympic Games. Buildings must be of two types (A and B), subject to the following conditions:

   Building A uses 180 units of wood, 60 units of concrete, has a capacity of 15 persons, takes 500 hours to build, and costs \$60,000.

   Building B uses 75 units of wood, 300 units of concrete, has a capacity of 24 persons, takes 600 hours to build, and costs \$100,000.

   At least 1100 units of wood and 3000 units of concrete must be used. How many buildings of each type must be constructed if the objective is to keep the cost of the construction at a minimum?

7. Use the information in Exercise 6, but the objective is to keep the time of construction at a minimum.

8. In a waste-treatment plant, there are three ingredients necessary for successful treatment. The minimum requirement per hour is listed in the table, along with the respective ingredients in each of the two products available for use and with the cost.

Product	Item A	Item B	Item C	Cost
Clene	8	6	1	.80
Puree	4	10	5	.65
Minimum requirement	24	30	18	

   Subject to the requirements, find the minimum cost for the ingredients.

Answers

2. $3W_1 + 5W_2 + 7W_3 \leq 3$

   $4W_1 + 2W_2 + 5W_3 \leq 2$

   Objective Function: Maximize $120W_1 + 100W_2 + 210W_3$.

4.

-1.2	1	0	.45	-.64	.09
1.28	0	1	-.18	.46	.36
28.8	0	0	7.2	31.8	85.4

   Solution: 7.2 of A and 31.8 of B; Minimum cost, \$85.40

6. 26 of building A and 5 of building B for a minimum cost of \$2,049,587.
8. 1.32 Clene and 3.36 Puree for a minimum cost of \$3.23.

CHAPTER 9 REVIEW EXERCISES

1. Find the solution set that maximizes profit; x and y are both greater than 0.

$$x + 2y \leq 16$$
$$x + y \leq 6$$

Profit: $10x + 3y$

2. Find the solution set that maximizes profit, x and y are both greater than 0.

$$\begin{aligned} x + 2y &\leq 16 \\ x + y &\leq 6 \end{aligned} \qquad \text{Profit:} \quad 3x + 10y$$

3. Find the nonnegative solutions that satisfy

$$\begin{aligned} 2x + 5y &\geq 10 \\ 4x + 3y &\geq 12 \end{aligned}$$

Minimize the cost function $3x + 4y$.

4. Find the nonnegative solutions that satisfy

$$\begin{aligned} 2x + 5y &\geq 10 \\ 4x + 3y &\geq 12 \end{aligned}$$

Minimize the cost function $5x + 3y$.

5. A review of past records indicates that Watertight Company can sell no more than 36,000 sleeping bags, 10,000 full-sized tents, and 20,000 pup tents per month. It has two production lines. The first assembly line can produce 5000 sleeping bags, 1000 full-sized tents, and 3000 pup tents per day. The second production line can produce 3000 sleeping bags, 1000 full-sized tents, and 1000 pup tents per day. The board has decided to produce no more than they know they can sell in a month. How many days per month should each of the two production lines work on manufacturing these three products? Watertight Company wishes to maximize profits. Suppose the profit on each sleeping bag is \$5, on each full-sized tent is \$14, and on each pup tent is \$10. What production schedule would you use to satisfy these constraints and maximize the profit?
6. The Par Golf Club Company manufactures two different types of golf bags, in which leather, aluminum, and zippers are used in the following amounts: For type I, 5, 4, and 3 units of the respective items are used; for type II, 6, 5, and 5 units of the respective items are used. They have available 560 units of leather, 460 units of aluminum, and 600 zippers. The Par Golf Club Company makes a profit of \$10 on a type I bag and \$15 on type II. Golf is a growing sport, and the demand for golf bags is unlimited. How many of type I and type II bags should be manufactured to maximize profit?
7. A company knows that they have to produce at least 25,000 gears and 14,500 axles. They have two machines that they can utilize to fill these orders. The first costs \$2000 per hour to operate and can produce 1000 gears and 300 axles per hour. The second costs \$2400 per hour to operate and can produce 600 gears and 600 axles per hour. How many hours should each machine operate and what is the minimum cost?
8. An appliance company has gotten a contract to supply at least 1,000,000 trash compactors and at least 1,000,000 garbage disposals to a mobile-home manufacturer. The company has two assembly lines to use to fill this particular order. Line 1 can produce 1000 trash compactors and 500 garbage disposals in an hour. Line 2 can produce 750 trash compactors and 1000 garbage disposals in an hour. It costs \$12,000 per hour to operate line 1 and \$15,000 per hour to operate line 2. How many hours should each line operate to minimize production costs, and what is the minimum production cost?

Chapter 9 Review Answers

2. $x = 6$, $y = 0$; maximum profit of \$60.
4. $x = 0$, $y = 4$; minimum cost of \$12.
6. Make 92 of type II for a profit of \$1380.
8. 400 hours of line I and 800 hours of line II with minimum cost of \$16,800,000.

Logarithms

CHAPTER 10

10.0 INTRODUCTION

Many of the topics of mathematics have remained constant throughout modern times. The theory of logarithms has been constant in mathematics, but the use of logarithms has changed dramatically in the last half of the twentieth century.

The predecessor of the handheld calculator in computational work was the slide rule. Computations on the slide rule are based on logarithms. The computer does most calculations as a series of additions and multiplications. However, the computer continues to utilize logarithms for noninteger powers and large exponents.

Logarithms are exponents and, therefore, are governed by the laws of exponents. Logarithms are essential in solving for exponential values. Most computers have a subroutine called up with some code such as log (x).

This chapter is designed to:

1. Provide the mechanics of translation between exponential and logarithmic form.
2. Provide a procedure to translate any number to its equivalent common logarithm.
3. Provide the properties of logarithms and illustrate how the properties are used to solve exponential equations.

10.1 DEFINITION OF LOGARITHMS

Exponential functions are used extensively in the scientific and business communities. In the chapter on sequences, geometric sequences involved exponents. Geometric sequences represented problems concerned with compound interest and half-life relationships. In exponential equation, the variable is an exponent: $a^x = b$ is an exponential equation, where x represents the variable. In many exponential equations, the value of the variable can be determined by a careful analysis of the relationship.

Example

$3^x = 9$, $3^x = 3^2$.

The bases are alike, therefore the exponents are equal and $x = 2$.

$$4^x = \tfrac{1}{16},\ 4^x = 4^{-2},\ x = -2$$

$$(\tfrac{1}{2})^x = .125,\ (\tfrac{1}{2})^x = \tfrac{1}{8}\ \ (\tfrac{1}{2})^x = (\tfrac{1}{2})^3,\ x = 3$$

These examples are carefully selected equations. The variables can be determined by knowing the laws of exponents and having some familiarity with numbers. Other exponential equations cannot be solved as readily.

Example

$3^x = 8$; the value of x is between 1 and 2 as $3^1 = 3$ and $3^2 = 9$.

$4^x = \frac{1}{10}$; the value of x is between -2 and -1 as $4^{-2} = \frac{1}{16}$ and $4^{-1} = \frac{1}{4}$.

$(\frac{1}{2})^x = .1$; the value of x is between 3 and 4 as $(\frac{1}{2})^3 = \frac{1}{8}$ and $(\frac{1}{2})^4 = \frac{1}{16}$.

In science and business, a value that is more precise is needed. To determine a more-precise value, a function called a *logarithm* is introduced.

Logarithm. A *logarithm* is an exponent to which a number is raised to yield a given number. In the expression $b^x = a$, x is the logarithm (or exponent), b is the base, and a is the given number.

Mathematically, $x = \log_b a$ is the way to write the definition of a logarithm. The logarithm is always equal to the exponent. The phrase $x = \log_b a$ is read "x is the logarithm of a to the base b" or "x is the logarithm to the base b of a."

Logarithms and exponents can have any numerical value. The base must be a positive number. If the given number is greater than 1, the logarithm is positive.

Example

$3^x = 9.9$; $9.9 > 1$, so the logarithm is greater than 0.

If the number is one, the logarithm is 0. If the number is between 0 and 1, the logarithm is negative. If the number is less than 0, the logarithm does not exist for the number.

To work with logarithms, it is necessary to translate from exponential to logarithmic notation and from logarithmic to exponential notation.

Example

$\log_2 8 = x$. Translate to exponential form, $2^x = 8$, so $x = 3$.

Example

$10^{-2} = .01$ is in exponential form. Since -2 is the exponent, the logarithm is -2. The base in the exponential equation is the base of the logarithmic statement. In this example, the base is 10. The value of the exponential form is the number whose log is being taken. Thus $\log_{10} .01 = -2$.

Example

$a^b = c$ is in exponential form. Since b is the exponent, the logarithm is b. The base of the exponent is a, so the base of the logarithmic statement is a. The value of the exponential form is c, so c is the value whose logarithm is being taken. Thus $\log_a c = b$.

EXERCISES

Write each of the following exponential equations as an equivalent logarithmic equation.

1. $5^x = 25$
2. $2^4 = 16$
3. $6^x = 216$
4. $10^x = .0001$
5. $1 = 7^x$
6. $a^0 = 1$

Write each of the following logarithmic equations in exponential form.

7. $\log_x 4 = 2$
8. $\log_x 2 = \frac{1}{2}$
9. $\log_2 x = -1$
10. $\log_3 3 = x$
11. $\log_{1/2} 16 = x$
12. $\log_{10} x = -3$

Find each logarithm.

13. $\log_2 16$
14. $\log_4 2$
15. $\log_4 32$
16. $\log_9 27$
17. $\log_8 \frac{1}{2}$
18. $\log_{10} .01$

Solve for x.

19. $\log_x 7 = 1$
20. $\log_8 2 = x$
21. $\log_x 81 = 4$
22. $\log_{10} (3x + 1) = 2$
23. $\log_2 (x - 2) = 3$
24. $\log_{10} x = -5$
25. $\log_2 x = -5$

Answers

2. $\log_2 16 = 4$
4. $\log_{10} .0001 = x$
6. $\log_a 1 = 0$
8. $x^{1/2} = 2$
10. $3^x = 3$
12. $10^{-3} = x$
14. $\frac{1}{2}$
16. $\frac{3}{2}$
18. -2
20. $x = \frac{1}{3}$
22. $x = 33$
24. $x = .00001$

10.2 COMMON LOGARITHMS

Any number greater than zero can be used as a base for logarithms. There are two bases used almost exclusively.

When the base of the logarithm is 10, the values of the logarithms are called *common logarithms.* In the statement $y = \log_{10} x$, y is a common logarithm. If the base is omitted from a logarithmic statement, the base 10 is implied. If any base other than 10 is used, it must be written. Base 10 logarithms are used extensively because 10 is the base of our decimal number system.

The other base used in logarithm notation is an irrational number called e, which has an approximate value of 2.7182818 . . . . The values of the logarithms with this base are called *natural logarithms.* The expression $y = \log_e x$ can be written as $y = \ln x$. Whenever ln is used as the abbreviation of logarithm, base e is implied. The subroutines of computers produce values of natural logarithms.

The values associated with common logarithms help give an understanding of concepts and uses of logarithms. It is extremely helpful to remember that logarithms are exponents. Common logarithms are exponents of 10.

Exponent form	Logarithm Form
$.0001 = 10^{-4}$	$\log .0001 = -4$
$.001 = 10^{-3}$	$\log .001 = -3$
$.01 = 10^{-2}$	$\log .01 = -2$
$.1 = 10^{-1}$	$\log .1 = -1$
$1 = 10^{0}$	$\log 1 = 0$
$10 = 10^{1}$	$\log 10 = 1$
$100 = 10^{2}$	$\log 100 = 2$
$1000 = 10^{3}$	$\log 1000 = 3$

If the logarithm of 500 is desired, the best the above table could provide is some value between 2 and 3. It is not 2.5. Tables are available that give the values of logarithms for all numbers. The log key on calculators also give values of common logarithms.

The table shows a small section of a common logarithm table.

x	0	1	2	3	4	5	6	7	8	9
4.5	.6532	.6542	.6551	.6561	.6571	.6580	.6590	.6599	.6609	.6618
4.6	.6628	.6637	.6646	.6656	.6665	.6675	.6684	.6693	.6702	.6712
4.7	.6721	.6730	.6739	.6749	.6758	.6767	.6776	.6785	.6794	.6803
4.8	.6812	.6821	.6830	.6839	.6848	.6857	.6866	.6875	.6884	.6893
4.9	.6902	.6911	.6920	.6928	.6937	.6946	.6955	.6964	.6972	.6981

The numbers on the extreme left are the first two digits of the number, and the column headings are the possible third digits of the numbers. This section of a logarithm table has three-place accuracy. It thus gives logarithms for three-digit numbers such as 456, .478, or 49,500 but not for 45,678 (without interpolation).

The logarithm of 4.67 is .6693. This means $10^{.6693} = 4.67$. In the table above, all the numbers except the leftmost column and the top row are logarithms.

Log tables begin for $x = 1.0$ and end for $x = 9.99$. To find the logarithm of a number outside of this range, it is necessary to write the number in scientific notation.

Example

Find the logarithm of 47,600.

Write 47,600 in scientific notation as 4.76×10^4.
From the logarithm table, log 4.76 = .6776, or $10^{.6776} = 4.76$.
Therefore, $4.76 \times 10^4 = 10^{.6776} \times 10^4 = 10^{4.6776}$.
The logarithm of 47,600 is 4.6776.

The procedure is outlined as follows.

Procedure for Finding the Logarithm of Any Number

1. Write the number in scientific notation.
2. Use the table of logarithms to find the logarithm of the decimal number. This is called the *mantissa.*
3. The exponent of 10 when the number is written in scientific notation is the other part of the logarithm, called the *characteristic.* In this example, 4 is the characteristic. The logarithm of 47,600 is .6776 + 4 = 4.6776.

Example

Find log 4,800,000.

$$4{,}800{,}000 = 4.8 \times 10^6$$
$$\log 4{,}800{,}000 = \log (4.8 \times 10^6)$$
$$= .6812 + 6 = 6.6812$$

Example

Find log .467.

$$.467 = 4.67 \times 10^{-1}$$
$$\log .467 = \log (4.67 \times 10^{-1}) = .6693 - 1 = -.3307$$

For numbers between 0 and 1, the logarithm can be written as a negative number, such as −.3307 or as a number that is part positive and part negative, .6693 − 1. This notation is useful for calculation. A calculator gives the remainder from the subtraction. For the above example, the display shows −.3307.

Finding the antilogarithm is the process of converting a logarithm or exponent to its decimal equivalent. The process reverses the process described above.

Procedure for Finding the Antilogarithm of Any Number

1. Divide the logarithm into its integer and decimal parts. The integer is the exponent of 10 in the scientific notation representation of the number. The decimal part is a logarithm. Find the decimal part in the body of the logarithm table and read the corresponding numbers from the left column and top row.
2. Convert the scientific notation representation to decimal notation. The result is the antilogarithm.

Example

Find antilog 3.6848.

$$\text{antilog } .6848 + 3 = 4.84 \times 10^3 = 4840$$

Example

$\log x = 1.6561.$
To find x, it is necessary to find the antilogarithm of 1.6561

$$\begin{aligned}\text{antilog } 1.6561 &= \text{antilog } .6561 + 1\\ &= 4.53 \times 10^1 = 45.3\end{aligned}$$

Example

$\log x = -2.3107.$
Because all the values in the logarithm table are positive, this negative number must be rewritten as the sum of a negative integer and a positive decimal:

$$-2.3107 = -3 + .6893$$

The next-smaller negative integer is used to determine the sum.

$$\begin{aligned}\text{antilog } -2.3107 &= \text{antilog } .6893 - 3\\ &= 4.89 \times 10^{-3} = .00489\end{aligned}$$

The properties of logarithms follow logically from the properties of exponents because logarithms are exponents.

1. $x^a \cdot x^b = x^{a+b}$ $\qquad \log (a \cdot b) = \log a + \log b$
2. $\dfrac{x^a}{x^b} = x^{a-b}$ $\qquad \log \left(\dfrac{a}{b}\right) = \log a - \log b$
3. $(x^a)^b = x^{ab}$ $\qquad \log a^b = b \log a$
4. $\sqrt[b]{x^a} = x^{a/b}$ $\qquad \log \sqrt[b]{x^a} = \log x^{a/b} = \dfrac{a}{b} \log x$

To demonstrate the meaning of the above properties, let log 2 = .3 and log 3 = .5. From these two values, find the values of the following:

$$\log 6 = \log 2 \cdot 3 = \log 2 + \log 3 = .3 + .5 = .8$$

$$\log 1.5 = \log \tfrac{3}{2} = \log 3 - \log 2 = .5 - .3 = .2$$

$$\log 12 = \log 3 \cdot 4 = \log 3 \cdot 2^2 = \log 3 + 2 \log 2 = .5 + (2)(.3) = 1.1$$

$$\log 27 = \log 3^3 = 3 \log 3 = 3(.5) = 1.5$$

$$\log \sqrt[3]{18} = \tfrac{1}{3} \log 18 = \tfrac{1}{3} (\log 9 \cdot 2) = \tfrac{1}{3} (\log 3^2 \cdot 2)$$

$$= \tfrac{1}{3} (2 \log 3 + \log 2)$$

$$= \tfrac{1}{3} (1.0 + .3) = .4333$$

EXERCISES

Use a three-place logarithm table to determine each value.

1. log 25.7
2. log .0257
3. log .531
4. log 53100
5. log 92.1
6. log .921
7. log .00719
8. log 719

Use a logarithm table to find x.

9. $\log x = 4.7210$
10. $\log x = 1.3784$
11. $\log x = -2.1469$
12. $\log x = -.2240$
13. $\log x = .1367$
14. $\log x = .9494$
15. $\log x = -4.2000$
16. $\log x = -1.2457$

Determine the following in terms of a and b, given $\log 3 = a$ and $\log 4 = b$.

17. log 12
18. log .75
19. log 16
20. log 27
21. $\log \frac{16}{9}$

22. $\log \frac{16}{27}$
23. $\log 2$
24. $\log \sqrt[3]{3}$
25. $\log \sqrt{12}$

Answers

2. -1.5901
4. 4.7251
6. -.0357
8. 2.8567
10. 23.9
12. .597
14. 8.90
16. .0568
18. $a - b$
20. $3a$
22. $2b - 3a$
24. $\dfrac{a}{3}$

10.3 LOGARITHMS AND COMPUTERS

There are two common bases for logarithms. The common logarithm utilizes base 10. The natural logarithm utilizes a base of 2.7182818 . . . . This number is referred to as e and can be generated from a series. Logarithm tables are not stored in a computer; the values are generated by a subroutine within the computer. The function log $(\frac{23}{7})$ generates a value of 1.189581. The natural logarithm of $\frac{23}{7}$ is 1.189581. If the common logarithm is needed, it can be determined by dividing the natural logarithm of the number by the natural logarithm of 10.

Example

Find the common logarithm of $(\frac{23}{7})$ on a computer.

$$\log\left(\tfrac{23}{7}\right) = \left[\log\left(\tfrac{23}{7}\right)\right]/\log(10)$$
$$= 1.1895841/2.3025851 = .5166298$$

The notation seems to be in conflict with the material presented in Section 10.2. The computer does not work in common logarithms. There is no need for two notations; the code to call the subroutine to generate natural logarithms is log.

When a program is written that involves exponents, the computer generates the result by brute force. For example, x^{19} is generated by multiplying x by itself the appropriate number of times. However, if x^{219} is needed, the computer automatically switches to logarithms to perform the calculations. The determination of the point of change from brute force to logarithms is part of the engineering of the computer system. This differs within computers. Whenever fractional exponents are used, the computer uses logarithms for the computation.

10.4 APPLICATIONS OF LOGARITHMS

Logarithms were useful for many years as tools for calculating tedious and impossible arithmetic operations, but today the calculator and the computer have made this use of logarithms impractical and unnecessary. However, logarithms are still needed to solve equations such as $3^x = 8$.

Example

Solve $3^x = 8$.

Apply the logarithm function to both sides of the equation: $\log 3^x = \log 8$. Apply the properties of logarithms to rewrite this as: $x \log 3 = \log 8$. Use the logarithm tables: $x(.4771) = .9031$. Use the techniques of solving equations to write: $x = .9031/.4771 = 1.893$.

Example

Find the seventh root of 200.

$$x = \sqrt[7]{200} = 200^{1/7}$$

$$\log x = \log 200^{1/7} = \tfrac{1}{7} \log 200 = \tfrac{1}{7}(2.3010) = .3287$$

$$\text{antilog}(\log x) = \text{antilog}(.3287)$$

$$x = 2.1316$$

Example

The formula for atmospheric pressure in pounds per square inch may be used to calculate the altitude:

$$p = 14.7(10)^{-0.09a}$$

where a is the altitude above sea level in miles. If the pressure is measured at 12.3 pounds per square inch, what is the altitude?

$$12.3 = 14.7(10)^{-0.09a}$$

$$\frac{12.3}{14.7} = .8367 = (10)^{-0.09a} \quad \text{Divide both sides by 14.7.}$$

$$\log(.8367) = \log(10)^{-0.09a} \qquad \text{Take the logarithm of both sides.}$$

$$-0.0774 = -0.09a\,(1)$$

$$.86 = a$$

The altitude is .86 miles.

EXERCISES

1. The number of bacteria in a culture is a function of time such that from one generation to the next, the number of bacteria doubles. This is the sort of information that gave us the geometric series 1, 2, 4, 8 . . . , or $a_n = 2^{n-1}$. Suppose we start with 100 bacteria and wish to find the number of generations that would be required to produce 1000 bacteria. The formula $f = b \times 2^n$ is used, where f is the final count, b is the beginning count, and n is the number of generations. Solve for n. If the time lapse for the growth from 100 to 1000 was 95 minutes, how much time is required to produce a generation of bacteria?
2. A certain radioactive element has a half-life of 34 years. How long will it take 2 grams of an element to become .1 of a gram if the formula $f = b(\frac{1}{2})^n$ holds, where f is the final amount, b is the beginning amount, and n is the number of 34-year cycles.
3. How long would it take for $10,000 to become $25,000 if the amount is compounded annually at the rate of 10%? The equation for compound interest is $A = P(1 + r)^t$, where A is the amount at the end of the period, P is the initial investment, r is annual interest rate divided by the number of compound periods in a year, and t is the number of years times the number of compound periods.
4. How long would it take for $20,000 to become $30,000 if the amount is compounded quarterly at an annual rate of 12%?
5. If $6^x = 200$, solve for x.
6. If $216^x = .456$, solve for x.
7. The annual deficit is doubling every n number of years, the deficit this year was $140 billion, and the deficit projected for next year is $150 billion. Solve for n in the formula $f = b(r^n)$, where f is the final deficit, b is the beginning deficit, n is the number of years, and r is annual rate of growth.
8. The population of the world is doubling every n years. The present population is 897 billion. The population last year was 843 billion. Solve for n using the formula from Exercise 7.
9. In how many years will $1000 be worth $2000 if it is invested at 9% compounded monthly? Use the compound interest formula in Exercise 3.
10. With annual inflation rate of 6%, how long will it take for $100 to be worth $50 in buying power?
11. Your rich grandmother put $10,000 in the bank on the day you were born. The savings book was found years later and you went to the bank to get your money. The amount you received was $25,000, and the money had been invested at 6%, com-

pounded quarterly. How old are you? Use the compound interest formula in Exercise 3.

Answers

2. $n = 4.32$ periods, or 147 years
4. $n = 13.7$ periods, or 3.4 years
6. $x = -.146$
8. $n = 10.05$ years
10. $n = 11$ years. Hint: $r = -.06$ because money is losing value.

CHAPTER 10 REVIEW EXERCISES

Solve for x.

1. $\log_2 8 = x$
2. $\log_4 .25 = x$
3. $\log .001 = x$
4. $\log 10{,}000 = x$
5. $\log_x 16 = 2$
6. $\log_x 81 = 3$
7. $\log_x 8 = 3$
8. $\log_x 5 = .5$
9. $\log (3x + 49) = 2$
10. $\log_4 (6 - x) = 3$
11. $\log_2 (x^2 + x) = 1$
12. $\log_3 (2x - 1) = 2$

In 13-16, let $\log 5 = a$ and $\log 2 = b$. Find each in terms of a and b.

13. $\log 20$
14. $\log 12.5$
15. $\log .4$
16. $\log \sqrt{10}$

Solve for x.

17. $25^x = 99.47$
18. $1.1^x = 4.45$
19. $243^x = 35$
20. $(.02)^x = .00336$

Chapter 10 Review Answers

2. $x = -1$
4. $x = 4$

6. $x = 3^{4/3} = 4.325$

8. $x = 25$

10. $x = -58$

12. $x = 5$

14. $2a - b$

16. $\frac{1}{2}(a + b)$

18. $x = 15.66$

20. $x = 1.46$

Four-Place Logarithms of Numbers from 1 to 10

APPENDIX

n	0	1	2	3	4	5	6	7	8	9
1.0	0.0000	0043	0086	0128	0170	0212	0253	0294	0334	0374
1.1	.0414	0453	0492	0531	0569	0607	0645	0682	0719	0755
1.2	.0792	0828	0864	0899	0934	0969	1004	1038	1072	1106
1.3	.1139	1173	1206	1239	1271	1303	1335	1367	1399	1430
1.4	.1461	1492	1523	1553	1584	1614	1644	1673	1703	1732
1.5	.1761	1790	1818	1847	1875	1903	1931	1959	1987	2014
1.6	.2041	2068	2095	2122	2148	2175	2201	2227	2253	2279
1.7	.2304	2330	2355	2380	2405	2430	2455	2480	2504	2529
1.8	.2553	2577	2601	2625	2648	2672	2695	2718	2742	2765
1.9	.2788	2810	2833	2856	2878	2900	2923	2945	2967	2989
2.0	.3010	3032	3054	3075	3096	3118	3139	3160	3181	3201
2.1	.3222	3243	3263	3284	3304	3324	3345	3365	3385	3404
2.2	.3424	3444	3464	3483	3502	3522	3541	3560	3579	3598
2.3	.3617	3636	3655	3674	3692	3711	3729	3747	3766	3784
2.4	.3802	3820	3838	3856	3874	3892	3909	3927	3945	3962
2.5	.3979	3997	4014	4031	4048	4065	4082	4099	4116	4133
2.6	.4150	4166	4183	4200	4216	4232	4249	4265	4281	4298
2.7	.4314	4330	4346	4362	4378	4393	4409	4425	4440	4456
2.8	.4472	4487	4502	4518	4533	4548	4564	4579	4594	4609
2.9	.4624	4639	4654	4669	4683	4698	4713	4728	4742	4757
3.0	.4771	4786	4800	4814	4829	4843	4857	4871	4886	4900
3.1	.4914	4928	4942	4955	4969	4983	4997	5011	5024	5038
3.2	.5051	5065	5079	5092	5105	5119	5132	5145	5159	5172
3.3	.5185	5198	5211	5224	5237	5250	5263	5276	5289	5302
3.4	.5315	5328	5340	5353	5366	5378	5391	5403	5416	5428
3.5	.5441	5453	5465	5478	5490	5502	5514	5527	5539	5551
3.6	.5563	5575	5587	5599	5611	5623	5635	5647	5658	5670
3.7	.5682	5694	5705	5717	5729	5740	5752	5763	5775	5786
3.8	.5798	5809	5821	5832	5843	5855	5866	5877	5888	5899
3.9	.5911	5922	5933	5944	5955	5966	5977	5988	5999	6010
4.0	.6021	6031	6042	6053	6064	6075	6085	6096	6107	6117
4.1	.6128	6138	6149	6160	6170	6180	6191	6201	6212	6222
4.2	.6232	6243	6253	6263	6274	6284	6294	6304	6314	6325
4.3	.6335	6345	6355	6365	6375	6385	6395	6405	6415	6425
4.4	.6435	6444	6454	6464	6474	6484	6493	6503	6513	6522
4.5	.6532	6542	6551	6561	6571	6580	6590	6599	6609	6618
4.6	.6628	6637	6646	6656	6665	6675	6684	6693	6702	6712
4.7	.6721	6730	6739	6749	6758	6767	6776	6785	6794	6803
4.8	.6812	6821	6830	6839	6848	6857	6866	6875	6884	6893
4.9	.6902	6911	6920	6928	6937	6946	6955	6964	6972	6981
5.0	.6990	6998	7007	7016	7024	7033	7042	7050	7059	7067
5.1	.7076	7084	7093	7101	7110	7118	7126	7135	7143	7152
5.2	.7160	7168	7177	7185	7193	7202	7210	7218	7226	7235
5.3	.7243	7251	7259	7267	7275	7284	7292	7300	7308	7316
5.4	.7324	7332	7340	7348	7356	7364	7372	7380	7388	7396

n	0	1	2	3	4	5	6	7	8	9
5.5	.7404	7412	7419	7427	7435	7443	7451	7459	7466	7474
5.6	.7482	7490	7497	7505	7513	7520	7528	7536	7543	7551
5.7	.7559	7566	7574	7582	7589	7597	7604	7612	7619	7627
5.8	.7634	7642	7649	7657	7664	7672	7679	7686	7694	7701
5.9	.7709	7716	7723	7731	7738	7745	7752	7760	7767	7774
6.0	.7782	7789	7796	7803	7810	7818	7825	7832	7839	7846
6.1	.7853	7860	7868	7875	7882	7889	7896	7903	7910	7917
6.2	.7924	7931	7938	7945	7952	7959	7966	7973	7980	7987
6.3	.7993	8000	8007	8014	8021	8028	8035	8041	8048	8055
6.4	.8062	8069	8075	8082	8089	8096	8102	8109	8116	8122
6.5	.8129	8136	8142	8149	8156	8162	8169	8176	8182	8189
6.6	.8195	8202	8209	8215	8222	8228	8235	8241	8248	8254
6.7	.8261	8267	8274	8280	8287	8293	8299	8306	8312	8319
6.8	.8325	8331	8338	8344	8351	8357	8363	8370	8376	8382
6.9	.8388	8395	8401	8407	8414	8420	8426	8432	8439	8445
7.0	.8451	8457	8463	8470	8476	8482	8488	8494	8500	8506
7.1	.8513	8519	8525	8531	8537	8543	8549	8555	8561	8567
7.2	.8573	8579	8585	8591	8597	8603	8609	8615	8621	8627
7.3	.8633	8639	8645	8651	8657	8663	8669	8675	8681	8686
7.4	.8692	8698	8704	8710	8716	8722	8727	8733	8739	8745
7.5	.8751	8756	8762	8768	8774	8779	8785	8791	8797	8802
7.6	.8808	8814	8820	8825	8831	8837	8842	8848	8854	8859
7.7	.8865	8871	8876	8882	8887	8893	8899	8904	8910	8915
7.8	.8921	8927	8932	8938	8943	8949	8954	8960	8965	8971
7.9	.8976	8982	8987	8993	8998	9004	9009	9015	9020	9025
8.0	.9031	9036	9042	9047	9053	9058	9063	9069	9074	9079
8.1	.9085	9090	9096	9101	9106	9112	9117	9122	9128	9133
8.2	.9138	9143	9149	9154	9159	9165	9170	9175	9180	9186
8.3	.9191	9196	9201	9206	9212	9217	9222	9227	9232	9238
8.4	.9243	9248	9253	9258	9263	9269	9274	9279	9284	9289
8.5	.9294	9299	9304	9309	9315	9320	9325	9330	9335	9340
8.6	.9345	9350	9355	9360	9365	9370	9375	9380	9385	9390
8.7	.9395	9400	9405	9410	9415	9420	9425	9430	9435	9440
8.8	.9445	9450	9455	9460	9465	9469	9474	9479	9484	9489
8.9	.9494	9499	9504	9509	9513	9518	9523	9528	9533	9538
9.0	.9542	9547	9552	9557	9562	9566	9571	9576	9581	9586
9.1	.9590	9595	9600	9605	9609	9614	9619	9624	9628	9633
9.2	.9638	9643	9647	9652	9657	9661	9666	9671	9675	9680
9.3	.9685	9689	9694	9699	9703	9708	9713	9717	9722	9727
9.4	.9731	9736	9741	9745	9750	9754	9759	9763	9768	9773
9.5	.9777	9782	9786	9791	9795	9800	9805	9809	9814	9818
9.6	.9823	9827	9832	9836	9841	9845	9850	9854	9859	9863
9.7	.9868	9872	9877	9881	9886	9890	9894	9899	9903	9908
9.8	.9912	9917	9921	9926	9930	9934	9939	9943	9948	9952
9.9	.9956	9961	9965	9969	9974	9978	9983	9987	9991	9996

Index

S

T

U

V

W

Z